Popular Mechanics

500 Simple Home Repair Solutions

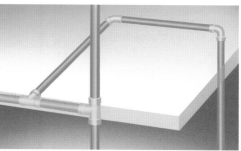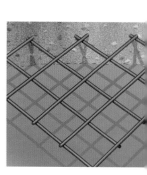

Popular Mechanics

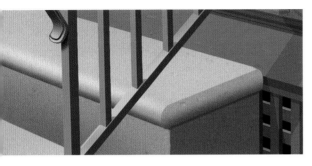

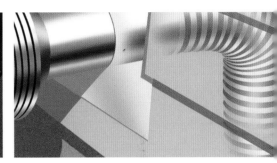

500 Simple Home Repair Solutions

Norman Becker, P.E.
author of the
Homeowners Clinic column

HEARST BOOKS
A division of Sterling Publishing Co., Inc.

New York / London
www.sterlingpublishing.com

Popular Mechanics 500 Simple Home Repair Solutions

Every effort has been made to ensure that all the information
in this book is accurate. However, due to differing conditions,
tools, and individual skills, the publisher cannot be responsible
for any injuries, losses, and/or other damages that may result
from the use of the information in this book.

Design: Barbara Chilenskas
Cover Design: Celia Fuller
Copy Editor: Bruce Macomber

Safety Note: Homes built prior to 1978 may have been
constructed with hazardous materials: lead and asbestos.
ou can test painted surfaces with a test kit available at most
hardware stores. Asbestos can be found in ceiling and wall
materials, joint compound, insulation, and flooring. Hire a
professional, licensed hazardous-removal company to check
for this and remove any hazardous materials found.

Library of Congress has cataloged the hardcover edition as
follows:
Becker, Norman.
  Popular mechanics 500 simple home repair solutions /
Norman Becker.
    p. cm.
  Includes bibliographical references and index.
  ISBN 1-58816-314-8
  1. Dwellings—Maintenance and repair—Miscellanea. I.
Title: 500 simple home repair solutions. II. Title: Five hundred
simple home repair solutions. III. Popular mechanics
(Chicago, Ill. : 1959) IV. Title.
  TH4817.3.B4223 2004
  643'.7—dc22
                              2004003887

10 9 8 7 6 5 4 3 2 1

First Paperback Edition 2008
Published by Hearst Books
A Division of Sterling Publishing Co., Inc.
387 Park Avenue South, New York, NY 10016

Popular Mechanics and Hearst Books are trademarks of
Hearst Communications, Inc.

www.popularmechanics.com

For information about custom editions, special sales,
premium and corporate purchases, please contact Sterling
Special Sales Department at 800-805-5489 or
specialsales@sterlingpublishing.com.

Distributed in Canada by Sterling Publishing
c/o Canadian Manda Group, 165 Dufferin Street
Toronto, Ontario, Canada M6K 3H6

Distributed in Australia by Capricorn Link (Australia) Pty. Ltd.
P.O. Box 704, Windsor, NSW 2756 Australia

Manufactured in China

Sterling ISBN 13: 978-1-58816-683-8
      ISBN 10: 1-58816-683-X

Contents

Foreword

Regardless if your house is brand new or "experienced", whether it's your first house or your fifth house, we know that you'll have questions about maintaining your home. That's why the "Homeowner's Clinic" column in *Popular Mechanics* magazine has been a reader favorite for 19 years, and why we've assembled here the questions and answers that are most on people's minds.

"How can I clean mildew off roof shingles? Should I be concerned about a bulge in a plaster ceiling? Our municipal water pressure is too low—can we boost it? What can I use to remove stains from a concrete walkway?" These and hundreds of other queries have poured in over the years, and now we've gathered the most frequently-asked into this easy-to-use guide.

What's your question? Is it about a wobbly garage door, or water in the basement? Do you have brick steps you wonder about painting? Or a bathroom exhaust fan that doesn't work well? Whatever your issue—whether it's in the house or around the yard—chances are good that it's covered here, with the reliable advice for which "Homeowner's Clinic" is known. It's easy to find the material that matters most to you.

Just glance at the "Contents" page, and you'll see how thoroughly we cover the areas and issues that come up again and again for homeowners just like you. From septic systems to ceiling fans, roof vents to retaining walls, you'll find just about every component and system in a home covered within these pages.

Even if you don't have any repair or maintenance questions right now, you will—and soon, as most homeowners can attest. And when those questions crop up, you'll be glad that *Popular Mechanics 500 Simple Home Repair Solutions* is right there with the answers.

—The Editors of *Popular Mechanics*

Introduction

As all homeowners know, a house is a complex structure. It is not maintenance-free, and over time something in the house or yard will invariably need repair or upgrading. This book will help answer the questions on the minds of homeowners. It is based on a the questions I have received from homeowners, over a 19-year period, as the writer of the Homeowners' Clinic column in Popular Mechanics magazine.

To maximize the usefulness of this book, it has been divided into three major categories: Exterior, Interior, and Electromechanicals. Each of the major categories is further subdivided into minor categories as shown below. And each of the minor categories is divided into subcategories, as shown in the Contents.

Exterior: Roof; Roof-mounted Structures; Paved Areas; Walls, Windows and Doors; Lot and Landscaping; Garage; Wood-destroying Insects

Interior: Attic; Interior Rooms; Basement—Crawl Space

Electromechanicals: Electrical System, Plumbing, Heating, Water Heaters, Air Conditioning

The questions I've received cover just about every component and system found in a house. For ease in finding answers to your questions, they have been placed in their respective subcategories. Even if you don't have any questions about your house at the present time, keep the book as a reference. You can be sure that questions will arise in the future.

Norman Becker, P. E.
Author, Homeowners' Clinic

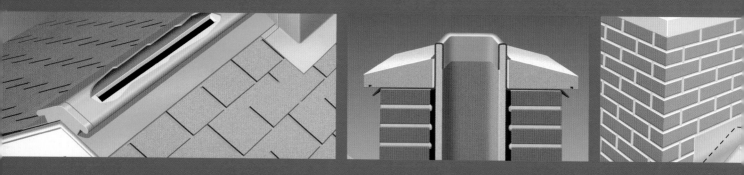

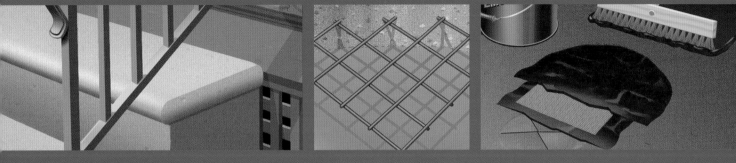

Exterior

1.Roof

SHINGLES • ROOFING PAPER (FELT) • SHINGLE MOSS,
MILDEW & ALGAE • ROOF VENTS • ICE DAMS • FLAT ROOF

SHINGLES

Roof Shingle Replacement

I want to have the roof of my 37-year-old ranch house reshingled. My roof has two layers of shingles, and I understand a third layer of shingles can be installed. I am hesitant to put the third layer on. My roofer feels confident that a third layer would be okay. I would appreciate your thoughts on this.

First, contact your local building department to find out whether a building permit is required and how many layers of roof shingles are allowed. Some building codes require a permit prior to reshingling a roof. Keep in mind that although general building codes may allow a maximum of three layers of shingles, your municipality may be more restrictive and allow only two layers.

Assuming that three layers of shingles are permitted, your decision on whether to strip off the old shingles or install the third layer over the existing shingles depends on the cost involved versus the aesthetics.

It's obvious that the job is considerably more expensive when the two existing layers of roofing are removed first. On the other hand, a roof with one layer of shingles is (sometimes) more attractive than a roof with three layers because the shingles lie down more smoothly when applied over a roof deck compared to when they are installed over shingles that are in poor condition.

Shingle Query

The shingles on our roof are worn, and we want to have the roof reshingled. We don't know whether to use asphalt or fiberglass shingles. Your opinion would be appreciated.

Either type of shingle will work for you. Your choice depends on aesthetics, availability and your budget. Generally, the more expensive shingles come with longer warranties, some of which can reach 20 to 25 years. Hire a reputable roofer who can supply references.

Many people, even roofers, confuse fiberglass and asphalt shingles. Fiberglass shingles are made with asphalt and should be referred to as fiberglass-asphalt shingles.

An asphalt shingle has a felt-base mat made from rags, paper and wood pulp. The mat is saturated and coated with asphalt, then surfaced with mineral aggregates. Fiberglass-asphalt shingles have a glass-fiber mat coated with asphalt and surfaced with mineral aggregates.

The difference between organic and fiberglass-based shingles is more of a concern to the roofer than to the homeowner. Fiberglass-based shingles were developed because roofers found that asphalt shingles, softened during hot-weather installations, were easily damaged. Fiberglass-based shingles are coated, not saturated, with asphalt and are not as easily damaged in hot weather.

However, in the northern United States, organic-mat shingles are often used. Fiberglass shingles are difficult to work with in very cold weather because they become brittle and can crack if flexed. Fiberglass shingles have a better fire rating than organic shingles. Nevertheless, the latter is acceptable.

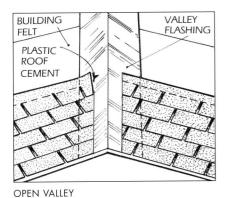

OPEN VALLEY

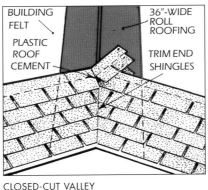

CLOSED-CUT VALLEY

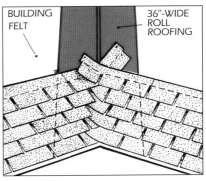

WOVEN VALLEY

Roof Valleys

Our 16-year-old, asphalt-strip-shingle roof will soon need replacement. What is the preferred way to handle roof valleys? In this area of New Jersey, roofing contractors do not use valley flashing but shingle continuously into and out of the valley. However, in the Midwest (St. Paul, Minnesota, area), contractors continue to use valley flashing (usually metal). Is there a preferred construction in treating roof valleys?

Without sounding facetious, the preferred way is the way the local roofers do it. There is no technical superiority between the styles. There are three basic types of valleys: open, closed-cut, and woven. The decision on the type of valley is usually the roofer's. That decision is based on past experience and aesthetics. If the valleys on all the roofs in the neighborhood are woven, most homeowners would be reluctant to have a roof installed with an open valley.

One reason for selecting one valley style may be due to how the local building department interprets national building codes. A roof with Class A (fire rated) shingles with an open valley may not be given a Class A rating, whereas it would be given a Class A fire rating with a woven valley.

Snowmelt in Roof Valley

We have a metal roof on our house constructed with a valley that collects the heavy snowfall. What can we do to encourage the snow to slide off? Can we put in heat tape or paint it with plastic?

I'm not familiar with a plastic that can be painted on a roof that will help. Heat tape, however, will work. The valley on a pitched roof has a shallower pitch than the sides, causing snow to drift in there. The tapes should be placed across the valley in an X or Z shape. It is important that the screws holding the tape-retaining device not be screwed into the bottom of the valley.

Double-layer-roof-shingling

My 25-year-old house needs reroofing, and I plan to do the job myself. I would like to lay new shingles over the old ones. However, the builder apparently skimped on the roof sheathing. Instead of the usual ½-inch-thick plywood, he used ⅜-inch sheets. Also, he

nailed the shingles directly to the sheathing without putting down building paper. Can I add a second layer of shingles? Or would this be too heavy for the 3/8-inch plywood, considering there can be a foot of snow on the roof in winter?

The builder didn't necessarily skimp on the roof sheathing. The required thickness depends on rafter spacing, the grade of the plywood, and the direction of its face grain. Look for the American Plywood Association (APA) stamp on the exposed underside of the plywood for the identification index. The index contains two numbers, such as 24/0 or 32/16. The left-hand number represents the maximum recommended spacing in inches between rafters when the face grain of the panel runs at right angles to the rafters. If the rafter spacing is not greater than 24 inches, your 3/8-inch sheathing is adequate.

I suggest you check with your local building department about a second layer. If your town allows only one layer you'll have to tear off the existing shingles.

The builder did skimp on the roofing paper. The paper serves as a backup water barrier between the shingles and sheathing if shingles are lifted, damaged or torn off by winds. The Asphalt Shingle Manufacturing Association, recommends using building paper. However, with a second layer of shingles, the absence of building paper is less critical.

Roof-shingle Replacement

Next spring I plan to have a new shingle roof put on my home. Should the old shingles be removed before the new roofing is installed? It's expensive to remove them, and I would like to leave them on if I can.

First, you should find out what, if any, rules apply in your area regarding this situation. Most municipalities allow a maximum of two layers of shingles, but some allow three. Check with your local building codes department for its requirements. If the number of layers of old shingles plus a new layer would exceed the limit, then the old shingles have to come off.

If there aren't too many layers, and they don't prevent the new shingles from being laid evenly, you can leave them on. However, if the old shingles are curled or cupped, the new shingles will not lay evenly over them and the old ones will have to be removed.

Loss of Shingle Granules

The roof of my house was reshingled four years ago during good, dry weather. Since reroofing, a uniform layer of granules about 1/8 inch thick has accumulated in the gutters and a significant amount has washed out of the gutters onto the ground. What level of granule loss is acceptable, and what effect does granule loss have on the shingles?

Excessive loss of granules from a relatively new roof is a serious problem because it will result in premature aging and deterioration of the shingles. The granules protect the shingles against the sun's ultraviolet rays and provide color.

Asphalt, which is used to impregnate the felt or fiberglass mat that forms the body of each shingle, provides the waterproofing property of the shingles. However, ultraviolet rays cause asphalt to degrade. If the asphalt is exposed to those rays because it lacks the protection of the granules, the shingles will deteriorate more rapidly than they would if the granules had remained intact.

Deterioration in shingles is marked by a loss of flexibility and an increase in brittleness. When this occurs, the shingles may curl away from the roof and crack, and pieces may break off. Aside from looking less attractive, the roof gradually loses its weather-resistant properties. Rain and melting snow may leak through the roof.

Some granule loss should be expected after installation, but an accumulation of $1/4$ inch from four-year-old shingles is excessive. You should contact the shingles manufacturer to file a warranty claim.

Aging Roof Shingles

My home has a 15-year-old, white, asphalt-shingled roof. There are terrible-looking black streaks forming on some areas. Short of waiting for a leak, how do I know when the shingles need to be replaced?

Whether shingles need to be replaced depends on their overall condition rather than the appearance of a leak.

Asphalt shingles are made by impregnating either organic felt mats or Fiberglas mats with asphalt and covering the top surface with mineral granules. The granules provide color and protect the shingles from being damaged by the ultraviolet rays of the sun. Most asphalt shingles have a projected life of 17 to 22 years. Longevity depends on shingle weight and exposure to the sun. Because southerly exposures get more sun, shingles on the south side will generally deteriorate more rapidly than those facing other directions.

As shingles age and weather, they become brittle and their corners curl. This makes them vulnerable to wind damage. Aging shingles also develop surface cracks and pitting due to granule loss. Continued granule loss exposes the mat, which leads to further deterioration and leaks.

Whether you should replace your shingles depends on their age and the extent to which they've deteriorated. You can always patch and coat a roof to extend its life. However, if about 30 percent of the shingles have deteriorated, I recommend that the roof be reshingled.

To Strip or Not To Strip

I'm thinking of putting new shingles on the roof of my home. Would I get as good and long-lasting a job by installing roofing paper and new shingles with longer nails over the old shingles, or is it necessary to strip all the old shingles and start from scratch?

The major factor to consider would be the number of layers of shingles already on the roof. Most municipalities allow two layers of shingles and some allow three. Check with your local building department to determine the code requirements for your area. Keep in mind that if the existing shingles are deteriorated, with curled, lifted, or eroded sections, you're likely to get a lumpy, uneven surface appearance. The old shingles must be in fairly good condition to provide you with the smooth surface you may be expecting.

When installed properly, there's no appreciable difference between applying a new roof over old shingles and stripping away the old shingles to start fresh. It's not necessary to install roofing paper when installing new shingles over the old roof.

Reroofing a Steeply Pitched Roof

I own a $2^1/2$-story home that needs reshingling, and I could use some tips on working safely on a multigabled, steeply pitched roof. I'm also wondering about the most economical and practical way to go about the job. The roof now has three layers of cedar shakes

over tar paper and boards of varying widths. Many of the boards have 1- or 2-inch spaces between them. We like the shakes, but the cost seems prohibitive compared to asphalt shingles.

To work safely on a roof, wear loose fitting clothes so you can move around freely. Also, wear soft-soled shoes to prevent slipping. High-top sneakers with good ankle support are recommended. Never go on a roof on a wet or windy day. Shingles can be slippery when wet.

On a steep roof the roofing ladder should be anchored in place with a bracket or framework that extends over the ridge. This is especially important for roofs with a 4° pitch or greater. Position the ladder so that you won't have to reach far out to work. If you do have to reach out, always hold the ladder with the other hand and keep your hips between the ladder rails.

As to reshingling, whether you use asphalt shingles or shakes you will have to remove the existing three layers first. If you use wood shingles, these can go on the existing boards. However, if you want to use asphalt, you'll have to cover the roofing boards with plywood sheathing to span the gaps between the boards.

You are concerned about the cost difference between shakes and asphalt shingles. Wood shingles are considerably more expensive than asphalt—as much as three or four times the cost,—and are more time-consuming to install. A professional might charge five to six times as much for the shakes.

Asphalt shingles come in various weight classes, from 210 to 400 pounds per roofing square. A roofing square is 100 square feet. The heavier the shingles, the greater the durability, and the higher the cost. The cost for top-of-the-line asphalt shingles, however, compares favorably with that of ordinary wood shingles.

Roofing under Solar Panels

I will soon need new shingles on the roof of my house. The problem is I have solar panels for my hot-water heater—three panels, each about 3 x 6—feet-mounted side by side in the center of the roof. They are held 6 inches off the surface by brackets bolted through the roof. Roofing contractors claim that since they can't get under the panels, they can't guarantee a watertight roof.

I've thought of two ways to solve the problem but, both have disadvantages. The first is to remove the panels while the roofing is being done. This involves unbolting the brackets, draining the antifreeze solution from the system, and unsoldering the copper pipes that carry the fluid in and out. Then the system would have to be recharged by a professional, since it must be done under pressure. The second method would be to box in the panels and effectively close off that part of the roof. I want to avoid this because a great deal of heat is generated under the panels, and I've read of roof fires caused by wood drying and igniting. I haven't been able to figure out any easy way to do this job.

There's no easy way to solve your problem. Apparently your solar panels were installed without considering the eventual need for reroofing. This is a problem of great concern to the roofing industry, and should also concern any homeowner contemplating the installation of solar panels. Panels should be mounted on brackets that provide sufficient clearance on the underside, and between the panels, to enable a roofer to install a new roof.

Either of the solutions will solve the problem. If you remove the panels, however, be sure to remount them on brackets that provide adequate strength as well as clearance. According to the U.S. Department of Housing and Urban Development, solar panels are subjected to uplifting force by wind striking the undersides. A panel array should be built to withstand wind gusts of at least 100 mph. These gusts result in an average wind load of 25 pounds per square foot on a sloped surface.

If you decide to box in the panels, strip the existing roof shingles down to the deck so the box/roof joint can be flashed in the conventional manner. Installing louvered vents on the sides will prevent excessive heat buildup under the panels.

Worn Shingles

The asphalt shingles on my 14-year-old house are turning up at the corners and have lost most of their white mineral surface. My roof is still serviceable, but I prefer an unweathered, white, white roof. Can the corners be made to lay flat, and can the old shingles be treated so they look white again?

When the corners curl and turn up, the roof deck and undersides of the shingles become vulnerable to water penetration by wind-driven rain. If you don't have too many curling shingles, resecure them with quick-setting asphalt roofing cement. Apply a dab about the size of a 25-cent piece under each corner, and press the shingles down so they lie flat. You may have to lay a weighted board over badly curled shingles. This repair should be done on a warm day when the shingles are pliable. Quick-setting roofing cement is available at hardware or roofing supply stores.

Widespread curling, however, indicates that the shingles have dried out and are losing their ability to keep the roof weather-tight. In this case, it's hardly worth your time trying to salvage them. The same holds true for your disappearing mineral granules. They are pressed into the asphalt coating of the shingles to provide a fire-resistant surface, color, and protection from the sun's ultraviolet rays. Granules of different colors are available to patch and match small areas, but since most of your shingles have lost their granules, you'll have to reshingle to get your white, white roof.

Cleaning Roof Shake

Please give us some advice on maintaining our 10-year-old shake roof.

The most important aspect of maintaining a shake roof is keeping it clean. Sweep and rinse out the slots between the shakes to keep leaves and debris from accumulating. These organic substances decay and trap moisture, contributing to the premature deterioration of the shakes. They can also prevent water from flowing off the roof, trapping it at the eaves, where it can back up and leak under the shakes.

Fortunately, cedar shakes resist rot because they contain natural preservative oils called extractives. However, shakes are not immune to rot. A homeowner in a humid climate, or whose house is in a heavily wooded area, should treat the shakes with a preservative containing a fungicide.

ROOFING PAPER (FELT)

Roofing paper: Yes or no

Ten years ago, a roofer replaced the roof shingles of my detached garage. Recently, while replacing two shingles damaged by a tree branch that fell, I noticed that there was no tarpaper under the shingles. Are there circumstances when omitting the tarpaper is permissible?

Roofing paper should be installed, but not every building code has a specific requirement for it. In fact, many municipalities don't even have building codes. Nonetheless, roofing paper increases the roof's fire resistance, provides secondary protection by shielding

ROOFING PAPER (FELT)

the roof deck from wind-driven rain, and can prevent a roof leak when shingles are torn loose in a storm.

A roofer might omit roofing paper to submit a lower bid, but it's worth paying a little more for it. Finally, not installing roofing paper may void the shingle manufacturer's warranty.

Roofing Paper

Recently, I was approached by a homeowner who was concerned that his roof did not appear to have roofing paper between the asphalt roof shingles and the roof deck. He was told by the contractor that the shingles were designed not to require roofing paper between the shingles and roof deck. Despite assurances by the contractor, the homeowner remains skeptical that any such shingle exists. I would appreciate your comments.

I checked with the Roofing Industry Educational Institute, and they said they knew of no shingle as you describe it. The type of shingle does not determine whether roofing paper is used, since the paper is supposed to provide additional protection against water penetration. Also, if a few shingles rip off in a windstorm, the paper is supposed to protect the roof deck. Most building codes require roofing paper.

Two building-industry trade groups disagree on the subject. The National Association of Home Builders says it is not necessary to use the paper, and the Asphalt Roofing Manufacturers Association says it is necessary.

According to NAHB, when many of the shingle companies closed their organic shingle plants and started making inorganic-based (fiberglass) shingles, they stopped making 15-pound felt (roofing paper) and started making fiberglass-based roofing paper. Contractors said the new paper wrinkled if it got rained on, or if it was left in the sun too long.

The Asphalt Roofing Manufacturers Association says roofing paper is necessary because it has a bearing on fire resistance. Shingles are fire-rated as a component

of a roof assembly that includes roofing paper and decking material. It also says the roofing paper helps shield the deck from wind-driven rain.

From the homeowner's point of view, I would recommend using roofing paper. It's worth it for the extra protection, and it may be required by the local building code.

Roofing Underlayment

In this area, 15-pound felt is applied to the roof deck prior to putting down shingles. Tyvek is tougher than felt and it works well on vertical sheathing. Can it be used instead of felt?

For readers unfamiliar with it, Tyvek is a tough plastic sheet applied over a house's wall sheathing prior to the installation of siding. It's made by DuPont. Tyvek was not designed as a roofing paper, and should not be used on a roof because it is too slippery. If applied on a pitched roof, it could cause bundles of shingles, or even a roofer, to slide off. It's probably correct to say that Tyvek is tougher, or more tear-resistant, than typical 15-pound building felt, but the point is moot. A good quality felt, properly installed, is more than tough enough to handle the rigors of having roofers walk on it, applying shingles.

Incidentally, a material similar to Tyvek is used in Europe. It differs from its American counterpart in that it is laminated with a material that makes it less slippery.

SHINGLE MOSS, MILDEW AND ALGAE

Dirty Roof

My home's roof is covered by asbestos-cement shingles that are 13 years old. Originally white, they are now dark gray with mildew. How should I clean them?

Aside from the mildew, the shingles' surfaces probably have been darkened by airborne dirt. One way to clean them is to use a solution of $2/3$ cup trisodium phosphate, $1/3$ cup of laundry detergent and one quart of household bleach. Add enough warm water to make one gallon of solution.

Wear rubber gloves to protect your hands and observe all safety precautions while working on the roof. Asbestos shingles are brittle, so be careful not to crack any.

Discolored Roofing

I have light-colored shingles on the roof of my house to reflect the summer sun, and thereby reduce my air-conditioning costs. The problem is that the shingles are discolored by dirty streaks of what I believe is mold. Is this discoloring due to mold, and is there anything that can be done to eliminate it?

The discoloration is probably caused by mildew spores. According to the Asphalt Roofing Manufacturers Association, this is a common problem and is often mistaken for soot or dirt. Fortunately, the mildew spores do not cause the shingles to deteriorate.

The discoloration cannot be eliminated, but it can be lightened temporarily with bleach. Gently sponge a dilute solution of chlorine bleach on the shingles

and then rinse it off with a hose. Don't scrub the shingles, or you will loosen their granules.

Shingle Fungus

I have a problem with fungus on my roof shingles. It has been building up for a few years now. The fungus, if that's what it is, is only on the front (north) part of the house. Is there something I can clean it with and is there any way I can stop it?

The discolorations on your roof are caused by microscopic algae spores that are carried by the wind, which, as they grow become visible. The algae stains can be removed by washing the roof shingles with a commercial-strength roof cleaner available at roofing supply stores or home centers. However, this is a temporary fix, as the algae will return. After cleaning the shingles, try the following for a more permanent solution. Install copper or zinc strips across the length of the roof and every few feet down the roof's slope. Rainwater flowing over these strips causes metal ions to leach out. As the water washes over the shingles, the ions inhibit the growth of the algae.

Controlling Shingle Fungus

What can be used to remove the black fungus from asphalt-hingle roofs without harming the shingles? The condition is common in this part of Florida.

This is a very common question, and one that we receive several times a year. For a temporary correction to the problem, you can treat the shingles with a solution of 50 parts water to 1 part liquid laundry bleach. Coat the shingles with a garden sprayer, working from the bottom of the roof to the top. Let the solution stand for a half-hour, then rinse the shingles with fresh water. Depending on how heavy the growth is, it may take two or three treatments.

A permanent solution to the problem is to install zinc or copper strips across the roof. The oxides that leach from the metal strips kill fungal growth.

If your shingles are approaching the end of their projected life you should consider installing new algae-resistant shingles. Although these shingles cost about $1 to $2 more per roofing square (100 sq. ft.) than regular shingles, they are guaranteed to remain free from any discoloration caused by algae, fungus, lichen, and cyanobacterial (blue-green algae) growth.

Algae Growth on Shingles

I would like to know what causes black streaks on asphalt shingles. How do I remove these streaks?

The black streaks are probably caused by microscopic algae spores that are carried on the wind and settle on the roof. The algae spores grow on the roof and become visible. Although algae grows best in a warm, humid environment such as is found in the South, it can be found throughout most of the states.

You can clean the roof using a mild solution of $1\frac{1}{2}$ ounces chlorine bleach per 1 gallon of water. Apply the bleach-water solution with a sponge or garden sprayer and then rinse it off with fresh water. Unfortunately, though, the algae discoloration will probably return.

To prevent this, you would need to install copper or zinc strips across the length of the roof. Place them immediately below the ridge and then every few feet down the roof's slope. Rainwater running over the strips forms a compound that washes over the shingles and retards algae growth.

Moss on Spanish Tile

What's the best way of removing moss from Spanish roof tiles? I'm concerned about damaging the tiles, or using a bleach solution and having it run off the roof and damage the paint on the side of the house.

The problem is how to clean Spanish tiles without breaking them. One tile manufacturer whom I spoke with said about 80 percent of a roof's tiles are not strong enough to support a person's weight. The manufacturer suggested hiring a pressure-washing contractor with a truck-mounted lift (a cherry picker).

Another option may be to clean the tiles from the edge of the roof while standing on a ladder. Use a pole-mounted sponge and a mild solution of detergent, bleach and water. Rinse the roof with a garden hose after you are done cleaning it. If any of the solution gets on the house, rinse it off as soon as possible.

Moss on Roofs

Our house, covered by asphalt shingles, is shaded by three large oak trees. In the winter rainy season, moss appears between the separations in adjacent shingles. A very thick growth of moss appears on the north and west sides of the roof. During a rain, moisture oozes up under the shingles lifted by the moss and leaks down into the house. Last summer I scraped off all the moss and glued the edges of the shingles down with roofing cement. However, this winter the moss quickly reformed and the roof is leaking again. Is there any way to keep the moss from reforming? I certainly do not want to cut down three beautiful old oak trees. If the house is reroofed, will the problem recur?

SHINGLE MOSS, MILDEW AND ALGAE

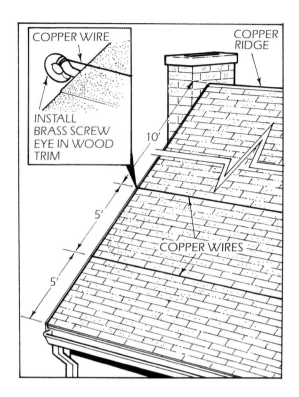

If moss exists as a thin layer on the shingles, it can be removed by wetting down the shingles with a copper sulfate solution ($\frac{1}{4}$ to $\frac{1}{2}$ ounce per 10 gallons of water) or a zinc chloride solution (10 percent zinc chloride and 90 percent water). Copper sulfate has a good residual effect.

The chemical can damage shrubbery and is corrosive to metal. Therefore, be careful with the overflow and make sure that gutters downspouts are clean before you start. After treatment, flush them thoroughly. When moss exists in clusters, it should be scraped off by hand, then treated as described above. Numerous chemical solutions are available in nurseries and garden supply stores for use on roofs to kill moss.

According to the Washington State Cooperative Extension, copper or galvanized ridges on a roof are often effective in keeping moss under control for about 10 feet down from the ridge. If a copper wire is stretched horizontally across a roof, the corrosive leaching of the copper should also provide moss control for about 4 to 6 feet down.

Controlling Roof Moss

Will stretching a copper wire across a roof help to control moss?

Yes, you can control moss by stretching bare copper wires horizontally about every 5 feet along the butts of the shingles. Also stretch a copper wire along the ridge. Rain and snow melt will carry the leachate from the copper wire down the roof. Copper leachate is corrosive to metal, so protect the inside of steel gutters with gutter paint.

Not only is moss on a roof ugly, but it can grow so vigorously that it raises and loosens shingles, making the roof vulnerable to water penetration during a driving rain or snow melt.

Another way to control moss is by spraying the roof with chemicals available in nurseries and garden supply stores. If you use these chemicals, follow the manufacturer's directions, because most of them are toxic to humans, corrosive to metals and harmful to plants.

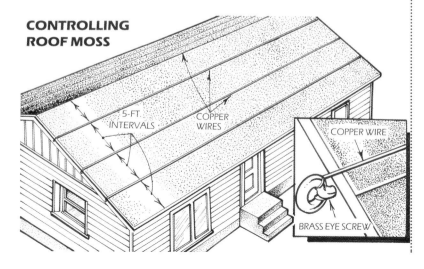

CONTROLLING ROOF MOSS

Moss Buildup on Roof

Over a period of 20 years, moss has grown and thickened on the tar-and-gravel-covered roof of my garage and on the asphalt shingles that cover my roof. At its thickest point, the moss layer is about 1½ inches. It doesn't appear to have damaged the roof, and the roof does not leak water. Should I remove the moss or leave it alone?

I recommend that you remove the moss layer before problems develop. Moss by itself will not damage the roof membrane or the shingles. However, it can trap airborne plant seeds, which may germinate and grow. During my years of doing home inspections, I have seen many tree seedlings growing out of roof moss and in gutters. As the seedlings grow, their root systems can lift the shingles or penetrate a weakened membrane joint. Also, roof framing is designed for a snow load, but not for a thick layer of moss. As the moss layer builds up, it introduces additional weight to the roof, and over the years it can cause a structural problem.

Roof-shingle Coating

The roof of my house is covered with asphalt shingles (235 pounds per square) that are 30 years old. Except for a small area with a southeast exposure, the roof is in excellent condition, due probably to the well-ventilated attic and the fact that the roof is covered with snow for three months of the year. I would like to extend the life of this roof by protecting it with a clear coating. Are there such products?

There are a number of asphalt-shingle roof coatings sold through roofing supply houses, home centers and hardware stores. The amount of life extension that these coatings would give the roof depends on the condition of the shingles at the time of application. Your shingles are already well beyond their projected life of 17 to 22 years. It is difficult to say whether the cost of applying the coating would be borne out by extended shingle life. However, the shingle coating may improve the look of the roof, at least for a short time.

ROOF VENTS

Ridge-vent Leakage

I recently renovated a home and installed a new roof with a ridge vent. However, I have a problem with wind-driven rain coming through the ridge vent and wetting the

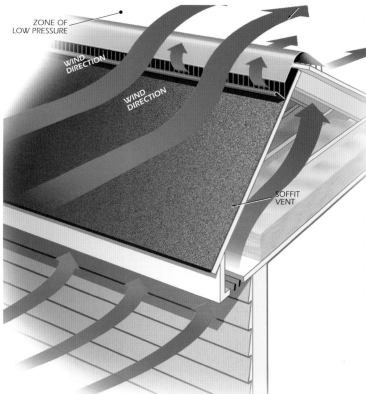

ZONE OF
LOW PRESSURE

WIND DIRECTION

WIND DIRECTION

SOFFIT VENT

ceiling in the room below the attic. How can this condition be corrected?

Your ridge vent may not be functioning properly. In order for it to work effectively, it must be accompanied by soffit vents, and its net-free ventilating area should be roughly equal to that of the soffit vents. The net-free vent area is the total vent area minus the area taken up by screens and louvers. Some vent manufacturers recommend 60 percent net-free vent area for the soffit and 40 percent for the ridge vent. Consult the manufacturer's instructions before installing a vent of any kind.

A properly balanced ventilation system works like this: Wind blowing over a roof creates a negative pressure at the roof ridge. Attic air is drawn out (exhausted) through the ridge vent, and makeup air is drawn in through the soffit vents. If the net-free area of the ridge and soffit vents is not balanced, there may not be sufficient air being exhausted to prevent wind-driven rain or snow from entering at the ridge.

Also, you should consider whether the ridge vent has exterior baffles. These baffles direct airflow up and over the vent, minimizing the possibility of wind-driven rain being blown directly in through the ridge vent. There are also ridge vents that have internal weather filters that provide additional protection against rain and snow entry.

Replacing the ridge vent with one that has both external baffles and an internal weather filter may solve the problem.

Ridge-vent Retrofit

My ranch home is about 20 years old. It has a 90-foot-long roof, and the width of the house is about 30 feet. The roof deck is in excellent condition, but the shingles need to be replaced soon. I was told that this would be the right time to retrofit the roof with a ridge vent, which would eliminate the need for running the attic fan for hours on a hot summer day. Can you advise me on the feasibility of this project and also whether this is a do-it-yourself job?

From your letter, I assume you have asphalt composition shingles on your roof. These shingles generally have a projected life of 17 to 22 years. Yes, this would be a good time to install a ridge vent (a low-profile, continuous louvered opening that runs almost the full length of the roof ridge). Actually, anytime, weather permitting, is the right time to retrofit a roof with a ridge vent. The installation is independent of the condition of the shingles. It is a do-it-yourself project providing that you observe the normal safety precautions when working on the roof.

A ridge vent is very effective in reducing the heat load during the summer months and minimizing the moisture buildup in the attic during the winter.

For maximum airflow through the attic, the ridge vent should work in conjunction with soffit vents. If you don't already have soffit vents, you should install them when you put in the ridge vent. You can buy ridge vents and soffit vents at home centers and lumberyards.

Venting a Roof with Unequal Pitch

I want to install a ridge vent on my house. On one side, the roof is steeply pitched, and on the other side the roof has almost no pitch. What kind of ridge vent will work in this situation? Also, there's a chimney in the center of the roof.

Normally, a ridge vent is installed on a roof with a pitch of 3-12 (a slope consisting of 3 inches of height for every 12 inches of horizontal distance) or greater. When installing the vent, a slot is cut through the roof

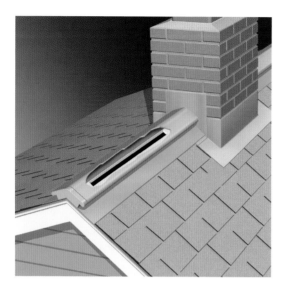

deck on both sides of the ridge, so both sides of the roof will be vented equally. In your case, since one side of the roof has a very low slope, the slot should be cut only on the steep slope.

Since the chimney is centrally located, you will need two separate slots. Each slot should be cut 12 inches from the chimney, and the vents should butt tightly against the chimney

In order for the ridge vents to be effective, they must work together with soffit vents. Air is supposed to flow in through the soffits and out through the ridge vent. If you don't have soffit vents, they will have to be installed. On the side of the roof with almost no slope, be sure there is an air path from the soffit to the slot on the opposite side of the ridge.

Roof Shingles Need Attic Ventilation

I live in a condominium complex, and I have been trying to get the board to pay attention to our roof ventilation. We had a new asphalt shingle roof installed a number of years ago.

Our undereave vents have smaller openings than necessary, and they are covered by a great amount of dust and paint that further hinders ventilation. I cannot convince the board that the higher heat of the attic space will shorten the life of the asphalt shingles, as the shingles' organic volatiles are driven off more rapidly. Do you have any definitive information regarding the necessity of keeping attic space cool to prolong the life of an asphalt shingle roof?

I agree with you that the heat from a poorly ventilated attic will result in accelerated aging of the asphalt roof shingles. I do not have any definitive information, nor do I know of any specific test, that proves it. However, my own experience of having inspected thousands of houses and having talked to many home inspectors is that a poorly ventilated attic will shorten the projected life of the shingles as well as result in an excessive moisture buildup in the attic. This also increases the heat load on the rooms below.

A roof-shingle manufacturer told me that adequate ventilation is required for the company's shingles to perform properly, and this is part of the shingles' application instructions. The manufacturer also mentioned that if the attic were inadequately ventilated, problems with the shingles would not be covered by warranty.

ICE DAMS

Ice-dam Prevention

Every winter we have extreme ice in the roof gutters, which results in water leaking on the wall inside the house. We would like to know if there is anything we can do outside (at the edge of the roof) to eliminate ice buildup.

Understanding the cause of ice dams will direct you to the cure. Ice dams are formed as a result of warm attic air heating the underside of the roof. This causes snow at the snow–shingle interface to melt, and the resulting water to run down under the snow. When the water reaches the edge of the roof, which is generally colder than the section above the living areas, it freezes, forming a dam that causes any additional melted snow to accumulate in a pool. Pitched roofs are designed to shed water, not to protect against standing water. As the melted snow accumulates, the water backs up under the shingles and leaks into the house. This causes damage that can appear in the form of soaked insulation; stained, cracked, and spalled plaster or drywall; damp and rotting wall cavities; damaged electrical components; and stained and blistered paint inside and outside the house.

The best method for preventing ice-dam problems is to keep the roof cold; that is, prevent the heat in the attic from heating the underside of the roof. This is done by heavily insulating the attic floor and ventilating the attic profusely. The latter can be done by installing continuous soffit and ridge vents. It is important that the insulation cover the top plate of the exterior stud walls. Care should be exercised when checking the insulation to ensure that it does not block the airflow

from the soffit vent to the ridge vent. Installing baffles over the insulation can do this. Baffles are available at home centers. In addition to the measures above, you should prevent heat from living spaces from leaking into the attic. A typical large heat leak is caused by an uninsulated attic stairway. Make a rigid hood from polystyrene-foam board insulation and use it to cover the hatch. Also when a house is built or reroofed, an ice-and-water shield membrane (rubberized polyethylene with an adhesive backing) should be installed from the lowest edge of the roof deck to at least 2 feet inside the wall line of the building. This prevents water leakage into the building should an ice dam develop.

Ice Dams

Many homes in the mid-Atlantic region experience property damage because of wall leaks resulting from thick ice in the gutters. I heard that some new homes have gutter heating elements. Are they available for existing homes?

The problem you refer to is called an ice dam. Ice dams begin when a layer of snow next to the roof melts and flows down to the roof edge. When this water freezes, a dam is created that causes further melted snow to accumulate in a pool. Pitched roofs are designed to shed water, not to protect against standing water. As the melted snow accumulates, the water backs up under the shingles and leaks into the house.

A fairly common practice for avoiding ice dam problems is to have the snow removed from the roof. In some areas, roofers do this work. Although this is a recurring expense, it may be preferable to annually repairing interior drywall and painting.

Another method is to use electrical heating tapes along the eaves, valleys, gutters and downspouts. This may not solve the problem completely. However, it is somewhat effective. The heating tapes (also known as deicing cable) reduce the ice dam buildup by creating heated channels that allow water to drain into gutters and downspouts.

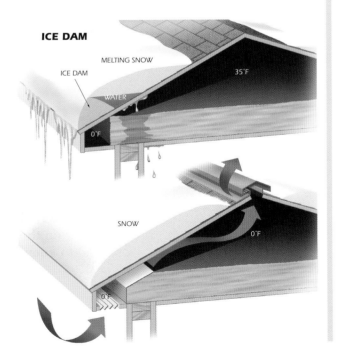

ICE DAM

MELTING SNOW

ICE DAM

WATER

35°F

0°F

SNOW

0°F

0°F

ICE DAMS

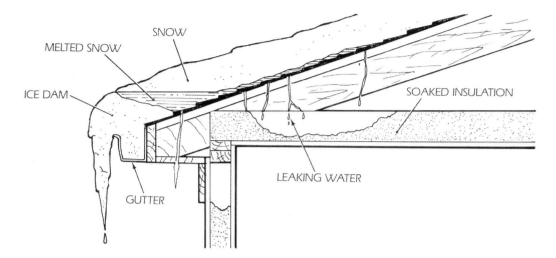

One method used when houses are built, or when they are reroofed, is to install a rubberized polyethylene membrane with an adhesive backing. It is installed from the edge of the roof deck to at least 2 feet inside the wall line of the building.

Ice-dam Woes

Every winter the north slope of my cathedral ceiling leaks when the snow starts melting. It doesn't leak in the rain. My ceiling gets ruined and has to be retaped and painted. Do you have any suggestions other than preventing or removing snow accumulation?

You have a classic case of leakage caused by an ice dam. Ice dams begin when a layer of snow next to the roof melts. When this water freezes, a dam is created which causes further melted snow to accumulate in a pool. Roofs are designed to shed water, not to protect against standing water, so the water penetrates the roof.

A fairly common practice for avoiding ice-dam problems is to have the snow removed from the roof. Although this is a recurring expense, it may be preferable to an annual retaping and painting. Another method is to use electrical heating tapes along the eaves and valleys. This may not solve the problem

completely, as its effectiveness is limited to the area where the tape is installed. Ice dams can still form farther up the roof.

The best method for minimizing an ice-dam problem is to maintain what's called a cold-roof. By over-insulating the ceilings and ventilating the attic profusely, the roof-deck temperature will be lowered to the point where the snow won't melt.

In a house with a cathedral ceiling, you normally cannot add more insulation under the roof deck, and, depending on the construction, you may not be able to provide additional ventilation. In this case, other than having the snow removed, I would try the heating tapes, which, although not perfect, may do the job.

FLAT ROOF

Blisters on Roll Roofing

A section of my old house has a flat roof, which is covered with roll roofing. The roofing has a couple of blisters that I would like to flatten. How should I repair them?

Blisters are fairly common on a flat roof. They are caused by water that has been trapped under the roofing. As

FLAT-ROOF REPAIR

ASPHALT ROOF EMULSION

ASPHALT ROOF EMULSION

ASPHALT ROOF CEMENT

ASPHALT ROOF CEMENT

FIBERGLASS MESH

BLISTER REPAIR

BLISTER REMOVAL

BLISTER

Cut an X into the blister across its perimeter. Fold the flaps back, exposing the area under the blister. Coat that area with asphalt roofing cement, and then fold the flaps down over the cement. There will be some overlap of the flaps, but try to keep them as flat as possible. Now coat the flaps with roofing cement and embed fiberglass mesh into it. In most cases, by the time blisters develop, the entire roof surface has weathered and is in need of a coating of asphalt emulsion. This helps extend the life of the roofing. In this case, after patching the blister, coat the roof surface according to the directions on the coating container,

the roof is heated by the sun, the trapped water expands. This raises the roofing and results in a blister.

With age and exposure to the sun, the oils in roll roofing materials evaporate, and the roofing becomes brittle. If you accidentally walk on a brittle blister, it will crack. Here is the way we recommend you repair a blister.

Paint a Rubber Roof

My rubber roof is several years old and I'm wondering whether there is a way to spruce it up. Can its appearance be rejuvenated with a coat of paint?

Rubber roof membranes can be painted with an exterior-grade water-based acrylic paint. Apply two coats. Don't use an oil-based paint on a rubber roof. It can damage the membrane by causing it to swell.

2. Roof-mounted Structures and Projections

CHIMNEY PROBLEMS • CHIMNEY FLASHING • CREOSOTE BUILDUP, LEAKAGE • GUTTERS/ DOWNSPOUTS • LIGHTNING PROTECTION

CHIMNEY PROBLEMS

Peeling Bricks

I have a problem with the bricks in the upper portion of my chimney. The faces on some of the bricks are peeling off—due, I'm told, to water being absorbed by the bricks and then freezing. As I understand it, the expansion of the ice will break off a brick's face if the brick has not been fired at a high enough temperature. The chimney is not falling down, but is there anything I can do, short of tearing down the chimney, to stop the bricks from peeling?

One of the main causes of spalling bricks near the top portion of a chimney is a defective masonry cap. Very often the cap is cracked, admitting water into the brickwork below. The water is absorbed by the masonry, and during cold weather it freezes. In the process, its volume expands, causing pieces of the brick face to come off. The condition can also result from cracked and open mortar joints between the bricks. The only way to control the problem is to stop the water from penetrating into the chimney. Damaged areas of mortar should be removed and fresh mortar installed.

Some people recommend applying a waterproofing agent, such as silicone, to the chimney to prevent or reduce water absorption. While we have received plenty of testimonials from readers over the years regarding the efficacy of this, we think it's unnecessary if the chimney is properly built, or as in this case, properly repaired.

It's possible, but not likely, that the chimney was built with bricks manufactured for interior, rather than exterior use. If this is the case, the bricks will continue to disintegrate. However, if the chimney was built by a mason, as opposed to an inexperienced do-it-yourselfer, it's unlikely that the wrong bricks were used.

My house is completely faced with brick and is about 28 years old. Each spring for the past four or five years, six to eight bricks on the upper portion of the chimney shed their outer surface, usually in one piece. I have also discovered some bricks around my closed-in patio with missing surfaces. Could this problem stem from freezing rain, snow, frost, or moisture? Can the faces of the bricks be glued back on? Can I prevent further peeling or shedding by applying a silicone water-sealant product?

The problem is indeed caused by the freezing of trapped water. Apparently, water is entering these structural elements through cracks or open joints.

You can't prevent the brick face from disintegrating by coating the area with water sealant.

In fact, the Brick Institute of America recommends against using silicones or other clear penetrating solutions on brick masonry structures because they often do more harm than good. Such sealants will generally not bridge cracks and open joints, and can trap moisture.

The only way to control the problem is to stop the water from penetrating by sealing cracks and open joints with mortar.

Gluing the faces back on the bricks won't last, eventually they will come loose again because they won't be

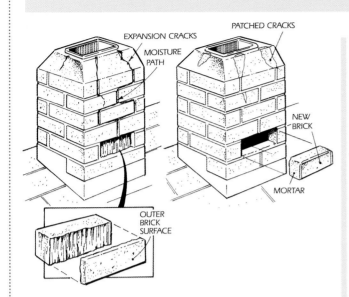

able to withstand weather changes. Your only option is to replace the deteriorated bricks.

To do this, chisel out the mortar that surrounds the affected bricks. It may be easier to break the bricks in order to remove them. Next, carefully chisel out the old mortar. Sweep dust and debris from the cavity and dampen the wall before the new bricks are installed to ensure a good bond. Mortar the appropriate surfaces of surrounding brickwork and the replacement brick. Center the replacement brick in the opening and push it in place. Remove excess mortar with a trowel. When the mortar becomes "thumbprint" hard, tool the joints to match the original profile.

Crumbling Mortar

I have a masonry chimney venting my oil-fired furnace. The mortar joints up to about 12 inches above the basement floor are beginning to crumble and turn to a white powder. How can I stop it?

The white powder is called efflorescence. It's caused by water passing through the masonry and absorbing soluble salts in the mortar. These salts are then deposited when the moisture evaporates.

The presence of efflorescence on an older chimney indicates that water is getting inside. It could be coming through cracks in the cement wash on top of the chimney, or in the bricks or mortar joints. To stop the efflorescence, you'll need to stop the water from seeping in.

You indicated that the mortar joints are beginning to crumble. When these joints are not in good repair, they permit water to penetrate the masonry. All cracks and deteriorated joints should be repointed. Repointing, sometimes called tuckpointing, is a labor-intensive job. When done with skill and technique, it will improve the water tightness of the chimney, enhance its appearance, and extend its life. The general procedure for repointing is as follows:

Remove the old mortar to a depth of at least 1/2 inch or until sound mortar is reached. You can do this by hand with a chisel, or with a power tool such as a tuckpointer's grinder. Unless you have experience repointing, use the grinder for horizontal joints only, and clean out the vertical ones with a chisel to avoid damaging the brick. When removing the mortar, avoid creating shallow or furrow shaped joints as they result in poor repointing. After the defective mortar is removed, clean the joints with a brush or, preferably, spray from a garden hose.

Carefully select the components of the mortar and proportion them correctly. Excessive shrinkage will reduce the mortar bond and make the joint more susceptible to moisture.

For best results, try to duplicate the proportions

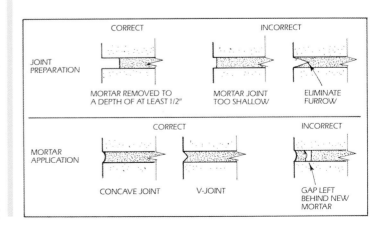

of the original mortar. However, if in doubt, use prehydrated Type N mortar. Mix one part portland cement, one part Type S hydrated lime, and six parts sand, proportioned by volume.

The mortar is generally applied to the cut joint with a tuckpointing trowel. The trowel should be narrower than the joints being filled in order to compact the mortar properly. Firm compaction is necessary to prevent voids and ensure a good bond with the brick and old mortar. For weathertight construction, tool all joints to either a concave or V shape.

Lime-bleeding Chimney

I have a beautiful fieldstone chimney, which is 10 years old. However, lime keeps bleeding from the chimney and staining the stones. I have cleaned the chimney using muriatic acid, but the lime comes back. Do you have any advice?

The white powdery deposits on the mortar joints are called efflorescence. The fact that you continually remove an efflorescence buildup indicates that water is working its way into the mortar joints either through cracks or open joints at the top or sides of the chimney.

Efflorescence is caused by mineral salts in the mortar that dissolve in water as water passes through the mortar joint. When the water evaporates from the surface of the joint, the salts are left behind.

Check the chimney closely. The top should slope away on all sides from around the flue to the chimney edge. The slope deflects rain and protects the joint between the flue and chimney. This sloped surface is vulnerable to cracking, and periodic sealing of this area is required. Other cracks and open joints should also be sealed.

Chimney-crown Leakage

We haven't been able to use our fireplace for five years because water leaks into the flue, hits the damper, then drips into the fireplace. We have had the roof reshingled twice around the chimney, caulked the metal flashing, and made a metal cap for the chimney. When it rains hard, we still get streams of water entering the firebox.

A defective or improperly installed chimney crown may be the cause of your problem. The crown is the sloped masonry at the top of the chimney. A common problem with the crown is that it is formed from mortar. The mason places mortar on top of the last course of bricks and slopes it away from the flue. The mortar shrinks

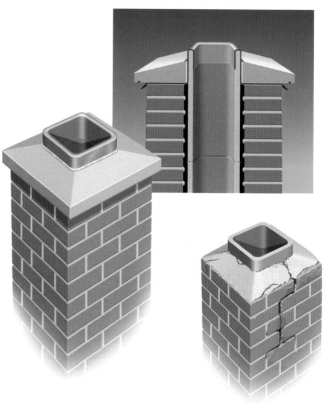

and cracks. Then gaps form between the crown and the flue. The cracks and gaps provide easy access for water to saturate the chimney.

The nature of a chimney is such that the flue liner gets warmer than the outer bricks and chimney crown. The flue will expand at a greater rate than the masonry surrounding it, so it's necessary to leave an airspace between it and the crown. To prevent water from entering this airspace, flexible sealant is installed between the flue and crown. The crown must also be protected with flashing. The flashing is installed so that it wraps from inside the flue, over the flue edge, behind the crown, and out the bottom of the crown. Any moisture that makes it past the sealed opening is diverted outside the chimney by the flashing.

Chimney Condensation

Condensation drips from the damper of my fireplace in the basement when we use the wood stove on the floor above. The stove and fireplace have separate flues. Both flues are contained in a large, open, brick chimney, and each is about 12 inches from the chimney's inside walls.

I have patched the chimney cap, bought metal covers for both flues, applied three coats of water sealer to the outside of the chimney, installed a dehumidifier in the fireplace, and removed a brick on each side of the chimney top to vent moisture. None of this has worked. Do you have any recommendations?

It sounds like you have a poorly built chimney. There shouldn't be 12 inches of open space between the flue tiles and the inner surface of the brick chimney. That area is usually filled with the rubble from the chimney's construction, such as broken bricks, stones, and mortar. The fill provides the flue tiles with lateral sta-

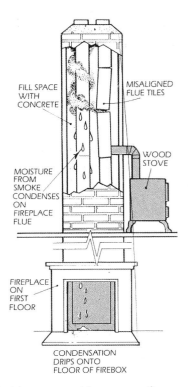

FILL SPACE WITH CONCRETE

MISALIGNED FLUE TILES

MOISTURE FROM SMOKE CONDENSES ON FIREPLACE FLUE

WOOD STOVE

FIREPLACE ON FIRST FLOOR

CONDENSATION DRIPS ONTO FLOOR OF FIREBOX

bility. Probably your problem stems from one or more open joints between the flue tiles for the wood-burning stove.

Chimney-flue tiles are generally butted one on top of another. A thin mortar bed at each butt joint holds the tiles together. However, it is possible that the tiles are not mortared together or the mortar joints have cracked apart. In that case, a chimney sweep's brushes may have jostled the tiles, opening several joints.

Wood smoke contains water vapor. Some of the smoke is probably seeping through the open joints into the area around the flues. Since the bricks and the fireplace flue are cool, the vapor in the smoke will condense on them and drip down to the fireplace.

To correct the problem, you need to fill the void around the flues with a loose cement mix. This will seal open joints and provide lateral stability. However, I suggest you hire a mason who specializes in fireplaces to correct the problem, because this procedure will considerably increase the weight on the chimney foundation, which may not have been designed to carry such a load.

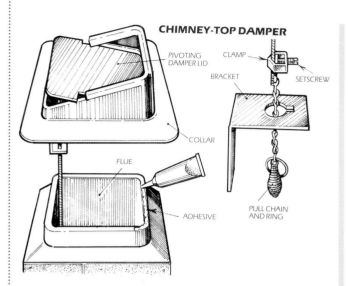

Chimney-top Damper

Our home, built in 1924, has a brick fireplace that has a lever-operated damper. The damper is jammed open. Can you advise us on how to fix this?

A good solution to your problem is to install a chimney-top damper. This spring-loaded damper is mounted over the flue opening at the top of the chimney. The device has a stainless-steel wire that runs from the damper down the flue into the opening of the fireplace. A pull chain and handle are attached to the wire with a setscrew and clamp.

To close the damper, pull on the wire and fasten the pull chain to a bracket at the top of the fireplace. To open the damper, release the tension on the wire.

Low-level Chimney

The top of my chimney (I have a gas furnace) is quite a bit below the level of the top of my roof. I have heard that this could be dangerous in the event of a moderate wind that could cause a backflow of gases into my home. I would appreciate any comments you might have on this subject.

The fact that the top of the chimney is below the top of the roof ridge is not necessarily a problem—as long as the chimney is terminated properly. The National Fire Protection Association (NFPA) Standard #211 states that chimneys for residential-type appliances must extend at least 3 feet above the highest point where they pass through the roof of a building, and must be at least 2 feet higher than any part of the roof within 10 feet measured horizontally. The measurements are made from the high side of the roof to the top of the chimney. This is sometimes referred to as the 3-feet, 2-feet, 10-feet rule. It applies to both masonry and metal chimneys, whether the chimney is connected to a central heating system, fireplace, or stove. You can extend the top of the chimney to provide additional height.

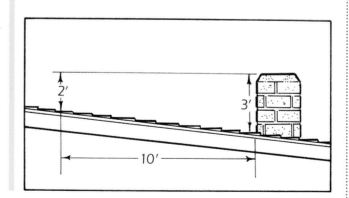

CHIMNEY FLASHING

Chimney Flashing

I'm writing in regard to a persistent water leak around my fireplace. All four sides of the fireplace chimney extend through the sloped roof. The chimney flashing has recently been replaced, but rainwater still leaks down the outside of the bricks inside the house. This occurs only after a steady or heavy rain. I'm not sure how the water is making its way inside the house and down the face of the fireplace. I would appreciate any help you can offer.

The leak is probably due to inadequate or faulty flashing between the chimney and the roof. Chimney flashing serves two functions: It seals the joint between the chimney and the roof to protect against rain or melting snow, and it accommodates any movement between the masonry and the roof.

Chimney flashing has four components. First is the base flashing on the lower face that covers the front and wraps around the chimney sides. Then comes the step flashing, which is nailed to the roof deck and bent up along the side of the chimney. This is followed by the cricket, behind the chimney. Last is the cap flashing, set into the chimney mortar joints and bent over the step flashing and the sides of the base flashing. Water runs off the cricket (and off the cap flashing) down the roof and over the step and base flashing.

Check the flashing at the chimney joints for the source of the leak. The flashing may not be installed properly, or it may be only partially installed. I have seen a number of chimneys without cap flashing, a somewhat common occurrence in reroofing jobs. In this case, asphalt cement is used to seal the joint between the step flashing and the chimney, but this frequently results in roof leaks, because water is able

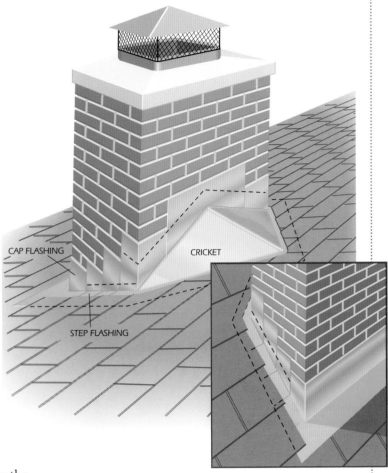

CAP FLASHING

CRICKET

STEP FLASHING

to penetrate at the point where the flashing meets the chimney.

If this is the case with your chimney, you can correct it by removing mortar and setting cap flashing into the joints. Then, tuckpoint mortar back into the joint and bend down the cap flashing. This is hard and skilled work, and in most cases it's best done by an experienced mason.

Water can also leak in if the chimney was installed without a cricket. Whenever the width of a chimney along the roof slope is more than 2 feet, a cricket (also called a saddle) should join the roof and chimney. The cricket prevents debris or snow and ice from piling up behind the chimney, which would cause rain or melting snow to back up under the shingles and leak into the house. The cricket also deflects water around the chimney and down the roof to the gutters.

CHIMNEY FLASHING

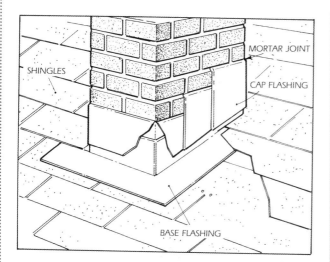

I have leaks on three sides of my chimney every time it rains. The roof was recently installed over a flat roof and is pitched at about 30 degrees. Metal flashing was installed around the chimney. After applying roof tar, the problem disappeared at one lower comer.

I've seen an advertisement for an aerosol spray product claiming to act as a water repellent for roofs. Do you know where I could get such a product and will it be effective in my situation?

I'm not familiar with the ad you refer to and I don't recommend using a water repellent spray to seal leaks around a chimney. Your leaks are probably due to problem with the flashing.

Proper chimney flashing is constructed to accommodate possible movement between the masonry and the roof. It's installed in two parts; base sections that are secured to the roof deck and extend up the side of the chimney, and cap sections which are secured to the chimney and overlap the base sections. If movement does occur, the cap sections slide over the base sections without affecting the water-shedding characteristics of the joint.

Check the flashing at the chimney joint for the source of the leak. The flashing may not be installed properly or only partially installed. I have seen a number of chimneys where the cap flashing was omitted entirely. Often, in this case, the base sections are sealed with asphalt cement in an attempt to make them watertight. This installation is "makeshift" and leakage readily occurs along the top edge where the flashing joins the mortar line.

If this is the case with your chimney, it can be corrected by setting metal cap flashing into the brickwork. Rake out a mortar joint to a depth of $1\frac{1}{2}$ inches, and insert the edge of the flashing in the cleared joint. Refill the joint with portland cement mortar. When the mortar is dry, bend the flashing down to cover the joint between the base flashing and the chimney to seal against water intrusion.

Chimney Cricket

We would appreciate your suggestions to correct a persistent leak in our roof by the chimney. This leak is causing water damage to the ceiling by the fireplace in our living room. My husband's efforts to tar the edges between the roof and chimney help for a while but don't correct the problem. We replaced the roof, but that made no difference. Will a sad-dle from the roof to the chimney help?

A cricket should solve the problem. Based on the photo that you sent, your chimney is about 4 ft. wide. Whenever the width of a chimney located along the slope of a roof is more than 2 feet, a cricket should join the roof and chimney. The cricket prevents debris or snow and ice from piling up behind the chimney. This can cause rain or melting snow to back up under the shingles and leak into the house. The cricket also deflects water running down the roof around the chimney.

The cricket's slope should be the same as that of the roof. If the cricket is large and exposed to view, it should be covered with the same shingles as those on the rest of the roof. A small or nonexposed cricket can be covered with metal. The joints between the cricket and chimney, and the cricket and roof, should be flashed.

CREOSOTE BUILDUP, LEAKAGE

Creosote Leakage

We live in a house built in 1800 and have a wood-burning stove. There is a problem with creosote leaking through the chimney joints. The chimney was recently taken down to the roofline and rebuilt, but creosote is still leaking from the chimney's mortar joints and into the attic. The chimney is cleaned yearly, so it is not a creosote buildup problem. How can we stop this?

The problem you describe is typical of an old, unlined chimney. A modern chimney has a flue liner made from fireclay tiles, each of which has a smooth, dense inner face that prevents creosote leakage. Creosote can still leak through deteriorated mortar joints between liner sections, however.

It's likely that your house's chimney is either not lined, or its liner has deteriorated. Creosote buildup is hazardous because it provides fuel for a chimney fire, and when it leaks out of the chimney it can pose a fire hazard by introducing flammable material into the house's wall and floor cavities.

You should have a certified chimney sweep check the integrity of your house's chimney. Ask the sweep about the options available for lining the chimney. A variety of metal liners are available. There is also a liner system in which a form is inserted in the chimney and inflated. Concrete is then poured around the liner. After the concrete hardens, the form is deflated and removed.

When selecting a chimney sweep, check to see if he or she is a member of the National Chimney Sweep Guild and certified by the Chimney Safety Institute of America (CSIA).

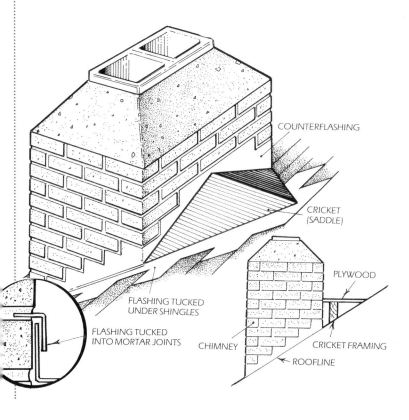

COUNTERFLASHING

CRICKET (SADDLE)

PLYWOOD

FLASHING TUCKED UNDER SHINGLES

FLASHING TUCKED INTO MORTAR JOINTS

CHIMNEY

CRICKET FRAMING

ROOFLINE

Crystallized-creosote Removal

Our house had a dangerous amount of creosote in its chimney when we bought it several years ago, and this has prevented us from using its fireplace. A number of chimney sweeps have tried unsuccessfully to remove it. Looking up from the fireplace, the creosote appears shiny, as if the flue were plastic lined. How can this be removed?

It was a wise decision not to use your fireplace; creosote buildup can result in a chimney fire. It's likely that the previous owner had a wood-burning stove connected to the fireplace. This often results in creosote buildup because the flue is sized for a fireplace and not a wood stove. The smoke condensed on the flue liner, and a thin layer of creosote was deposited each time the stove was used. This can quickly cause a buildup of creosote that is impossible to remove by simply sweeping the flue.

A chimney sweep can remove the creosote using a rotary chain tool attached to a drill. The device is lowered down the flue by pole sections, and the spinning chain scours away.

A chimney sweep with whom I spoke said the process removes 75 to 80 percent of the buildup.

When selecting a chimney sweep, check that he or she is a member of the National Chimney Sweep Guild and has been certified by the Chimney Safety Institute of America (CSIA).

Chimney Leaks Creosote

A brick chimney extends up between two unheated rooms in our attic. For the past 35 years, it appears that creosote has been leaking out of the chimney and staining the wallpaper. Is this dangerous? And if it needs repair, what kind of technician handles this?

The problem is dangerous, and it should be fixed as soon as possible. The leaking creosote indicates that there are openings in the chimney wall that extend to the flue. Incidentally, the black stains you see may not be creosote. Creosote is formed from the incomplete combustion of wood or coal. Instead, the material may be a sooty, oily film resulting from the incomplete combustion of fuel oil.

Regardless, both creosote and fuel-oil film have corrosive elements. As flue gas rises, it cools to the point that a corrosive condensate may form and adhere to the chimney liner. This corrosive material attacks the flue lining and its mortar joints. If there's no flue lining, the corrosive deposits attack the brick and mortar.

Eventually, cracks will develop and flue gas will escape.

When the chimney is inside the house, rather than outside, flue gases can seep through the cracks and into the attic or living areas. This is a fire hazard if there's wood framing near the chimney, and it's a health hazard: Flue gas contains carbon monoxide. The safest solution is to install a new flue lining. Many chimney sweep companies do this work.

Creosote Problem

When I was installing a wood stove in my fireplace, I discovered that the smoke chamber above the damper was covered with creosote. The damper opening is too small to reach up into this space from below. A chimney cleaning brush pushed down from above will not reach the area. Do you have any suggestions?

If you can't remove the damper plate (almost all damper plates can be removed or pushed aside once the handle is removed), then do the job with a "pull-cord" chimney brush, which has a cord attached at its top and bottom. One person stands on the roof, and, while holding one end of the cord, drops the brush down to a person by the fireplace hearth.

Make sure to drop enough cord down so the brush is positioned in the smoke chamber. Then, alternately pull from the top and bottom to brush away the creosote.

GUTTERS/DOWN-SPOUTS

Are Gutters Necessary?

Are rain gutters absolutely necessary for a home? What happens if you don't install them? Are there advantages over having them to not having them?

Gutters and their associated downspouts (also called leaders) are installed on a structure to control and direct rain runoff from the roof. The absence of gutters might result in water seepage into the basement or crawlspace, rotting sections of wood trim, damage to foundation plantings and the erosion of topsoil adjacent to the house. Whether they are masonry constructed or have long, overhanging eaves, most houses that are not in the Snow Belt would benefit from gutters. In the Snow Belt, gutters are considered a nuisance and require lots of maintenance; often snow and ice will tear them from their supports.

Downspout Dry Wells

I'd like information on downspout dry wells. For instance: What is the daily rate, in liters, that can be passed through one? Is it possible to use more than one drum at a time?

The amount of water any dry well can accept depends on the size of the well and the percolation rate of the soil. A hydrologist can determine the amount of water entering the dry well by taking into account the rate of the rainfall, the roof area, the size of the gutter channel and the cross-sectional area of the downspout. However, most homeowners are realists and not theoreticians. If one dry well is not effective, they put in a second and a third.

Dry wells can be installed in series or in parallel. When installed in series, each well has an outlet near its top. If the dry well fills up before the water can leach into the ground, the water flows through the outlet to the next well. When dry wells are installed in parallel, a distribution box—similar to that used in a septic system—is

installed. The box, which is plastic or masonry, has an inlet from the downspout and two or three outlets, each going to a dry well. The dry wells should be spaced so that they have enough soil around them to absorb all the water they will receive.

I can't find plans or information on building a dry well for my downspouts. I have looked in "how-to" books and magazines. I would appreciate any information that you can supply.

A dry well is often just a large hole in the ground covered with boards and filled with rocks to keep the sides from collapsing. A better dry well, however, is a clean steel or plastic drum buried 18 inches below grade, at least 10 feet from the house's foundation. Do not bury the drum closer, because runoff seeping from it could enter the house's basement through cracks in the foundation.

Drill numerous holes through the drum with a 1/2-inch-diameter drill bit, and fill the drum with medium-size stones. Make the drum's lid from pressure-treated lumber or a thin slab of concrete reinforced with wire. This design is better than a rock filled hole because it is less likely to fill with silt eroded from the soil.

A dry well should work in most situations because its large surface area allows water to percolate into the

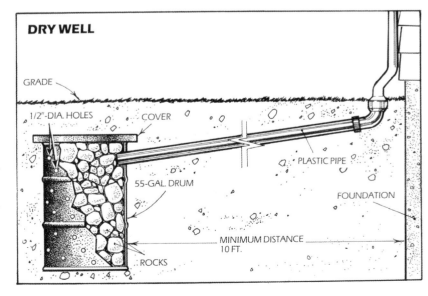

DRY WELL

GRADE

1/2"-DIA. HOLES   COVER

55-GAL. DRUM

PLASTIC PIPE

FOUNDATION

MINIMUM DISTANCE 10 FT.

ROCKS

ground before the drum is filled with water. However, if your area has high ground-water levels in some seasons, or year round, there will be times when the drum is filled with water and can't accept runoff. Also, a dry well can become ineffective if it fills with debris that washes out of a house's gutters, so keep your gutters clean.

Gutter Clogging

My house has standard aluminum gutters and downspouts. The gutters are constantly getting clogged. On a recent trip to New Hampshire, I noticed that practically none of the houses had gutters. Instead, they had what looked like sheathing extending up the roof for about 3-feet. Would you know the principle behind this system?

What you saw was not the solution to clogged gutters or keeping gutters clean. In the Snow Belt, many houses don't have gutters because they contribute to ice damming or get torn loose by snow sliding off the roof. The sheetmetal strip is slippery when wet and enables the snow at the lower edge of the roof to slide off easily, rather than build up into an ice dam, which can cause a backup of water under the shingles and into the house.

The sheetmetal is generally extended far enough up the roof to cover a point at least 12 inches inside the interior wall line of the building.

LIGHTNING PROTECTION

Lightning-Protection System

My house, built on the shore of a man-made lake in Venice, Florida, was recently struck by lightning. I have been told that this is the lightning capital of the world, yet I can't seem to find lightning-protection equipment or information. Can you tell me how to install this type of protection system?

I don't know about the world, but you do live in the lightning capital of the United States.

Although you may be handy, installing a lightning-protection system is not considered a do-it-yourself project. A lightning strike can cause considerable injury and property damage if the protection system is installed improperly.

There are many details you have to be familiar with to install a lightning protection system. Here's one

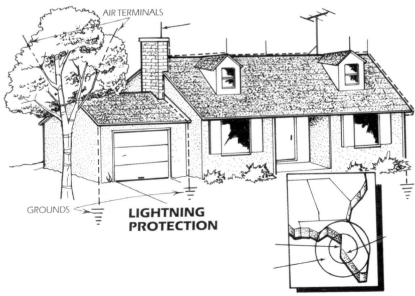

AIR TERMINALS

GROUNDS **LIGHTNING PROTECTION**

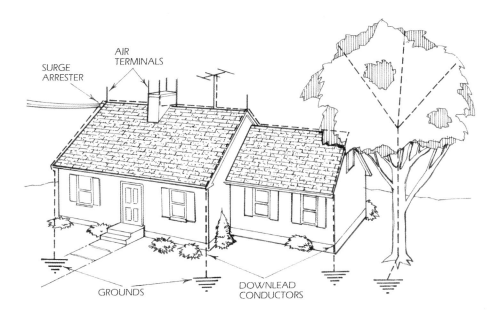

example of how complicated this can be: Although copper air terminals are acceptable on most sections of a roof, a terminal within 2 feet of a chimney top must be protected from corrosive flue gases by a covering of hot-dipped lead.

To ensure that a lightning-protection system complies with nationally recognized codes and standards, the installation should be performed by a UL-listed installer.

Lightning Damage

I am building a home in a mountainous area of California. Recently, a nearby cabin was struck by lightning and suffered considerable damage. How can I protect my new home from lightning damage?

Although there are about 90 million lightning strikes each year in the United States, most homes don't have lightning protection. Homeowners and builders determine whether a lightning-protection system is necessary by weighing the probability of a strike against the cost of a system.

Because of their location, some homes have a higher risk of lightning damage than others. If your house will be in a high-risk area, then it's wise to install a lightning-protection system. These systems have two objectives. They must provide a direct path for a lightning bolt to follow to ground, and they must prevent property damage and personal injury as the bolt travels this path to ground.

It's important that the person or company that installs the lightning-protection system be listed by Underwriters Laboratories (UL). This ensures that the installation will comply with current nationally recognized codes. After the system is installed, it should be inspected by UL's Lightning Protection Division.

3. Paved Areas Around Structure

DRIVEWAY • PATIO, WALKWAYS AND STEPS

DRIVEWAY

Sealing a Driveway

The surface of my asphalt driveway is drying out and beginning to loosen and ravel. I would like to apply a sealer but do not know which is best: the coal-tar-based products, which require a lot of stirring, or the new, plastic poly-based, which are ready to use right out of the can. Any advice will be greatly appreciated.

Coal-tar sealer requires a lot of stirring because it contains sand, which provides traction and acts as a filler. When a bucket of coal-tar sealer has been sitting on the shelf for a while, the sand settles to the bottom. The sealer needs to be thoroughly stirred to bring the sand back into suspension in the sealer. This product is more effective then the other on an inclined driveway during wet or slippery conditions because the sand provides traction. On a level driveway they are both effective.

If you want to extend the life of your driveway, don't wait until it is drying out and beginning to loosen. Driveways should be sealed yearly if done by homeowners, but if they are professionally sealed, then the interval can be extended to two to three years.

My asphalt driveway is five years old, and people have told me that I should have a sealer put on it. Others say that it's the wrong thing to do. What is your opinion about sealing a driveway?

Whether or not you should seal your driveway depends on its condition. If it exhibits extensive cracking, then sealing it will not prolong its life because the sealer will not bridge the cracks, it will only change the appearance of the driveway. On the other hand, as long as the driveway surface is in reasonably good condition, it should be sealed every one to three years, depending on the severity of the climate. Small cracks should be cleaned out with a wire brush or compressed air and then filled with a crack-filling material. Then the driveway should be sealed.

It's best to seal driveways when they are less than a year old, although they can be sealed successfully years later. Bear in mind that a new asphalt driveway is flexible and elastic. It will expand and contract without cracking. As a driveway ages, however, the pavement loses its elasticity and becomes brittle and vulnerable to cracking.

When water enters cracks in asphalt, it works its way down to the base below, reducing the base's compressive strength. This can cause depressions to form and, eventually, potholes. If the water freezes, the ice can increase the size of the cracks.

Also, a coat of sealer protects the driveway from damage caused by the sun's rays, and if the sealer is a coal-tar emulsion, it will protect the driveway from surface unraveling caused by oil drips or gasoline spills. In cold climates, sand is added to the sealer to provide slip resistance.

If a contractor is going to seal your driveway, do the following: Check to see if that person is licensed, and call the local consumer protection office to find out if there are any complaints against that contractor. Ask the contractor what proportion of the sealer is coal tar. It should be about 70 percent coal tar and 30 percent water. Typically, the 5-gallon drums of coal-tar driveway sealer that a homeowner buys at a home center or hardware store are 50 to 70 percent water and 30 to 40 percent coal tar. Lastly, ask if the sealing is done by hand or is sprayed on. A sprayed coat of sealer is more uniform in its thickness and its appearance. Sealer applied by brush, roller, or squeegee may exhibit some surface imperfections, such as brush marks or ridges.

Crumbling Concrete

What causes a section of concrete driveway to fail after only eight years? The apron—between the street and sidewalk—remains strong, without any pitting at all. However, the section between the sidewalk and house is crumbling to stone, sand, and powder. Both sections were poured the same day, but from different truckloads. Both are exposed to the same salt from city streets, but I applied no salt to either section myself.

It is possible that the concrete mix in the trucks was different, even though the mix came from the same supplier. Concrete with too much water and/or not enough cement is likely to be too weak to withstand an automobile load. You don't mention whether or not the deteriorating driveway section collects water runoff from a side yard or downspouts. Water that puddles and then freezes and thaws can shorten the lifespan of concrete.

If you park your car on this section, it's also possible that the deicing salts from the city streets that accumulate on the undercarriage drip off onto the driveway, causing surface deterioration. According to the American Concrete Institute, deicers containing ammonium salts such as sulfate, nitrate or chloride should, as much as possible, be avoided on concrete.

Since the deteriorated section is beyond rehabilitation, it will have to be replaced. This time, make sure the concrete mix contains an air-entraining admixture, which is required if the concrete is exposed to freezing and thawing.

Air-entrained concrete contains millions of microscopic air bubbles per cubic foot. These air bubbles act as relief valves, because they provide tiny cavities for the expansion of water when it freezes. This type of concrete is highly resistant to chemical deicers and the deteriorating action of the freeze–thaw cycle.

Paving Over Concrete

I have a concrete driveway that is somewhere between 20 and 30 years old. I have patched some of the worst cracks, but this is very difficult work. Is it possible to cover the driveway with asphalt pavement? If so, how do I prepare the larger cracks ($\frac{1}{2}$ to 1 inch wide) to receive the pavement overlay? I hope you can advise me.

Your concrete driveway can be covered with asphalt pavement. However, the new pavement will eventually experience reflection cracks (fissures that reflect the cracks in the pavement below). These cracks are caused by vertical or horizontal movement in the substrate pavement, and they are brought on by expansion and contraction with temperature or moisture changes. The only way to avoid reflection cracking is to break the substrate pavement into rubble or remove it entirely and prepare a new base.

Prior to paving, the driveway must be cleaned and its cracks sandblasted or wire brushed. In addition, the cracks must be filled with a flexible sealant. It's important that the contractor chooses an appropriate sealant. There are many sealants to choose from, and there are significant cost and performance differences among them. Hot-applied asphalt sealant is the most effective material in terms of cost and mechanical characteristics. However, this sealant must be professionally applied.

Heated Driveway

I recently bought an older home that has a 20-foot-wide × 100-foot-long gravel driveway. I plan on having either a concrete or an asphalt driveway installed, and I would like the new driveway to incorporate some type of heating system to melt snow and ice. Can you recommend a do-it-yourself system that would not be too expensive to install or operate? It would have to be durable and maintenance-free.

There are two types of built-in systems for melting snow on a driveway—electric and hydronic. Both are expensive to install and neither is a do-it-yourself project. Because of the equipment involved in a hydronic system, the installation cost is greater than an electric system. However, the operating cost for an electric system is generally greater than that for a hydronic system.

With regard to trouble-free service, both systems will normally provide trouble-free service as long as the asphalt or concrete driveway is not disturbed or damaged.

The tubing in a hydronic system may be constructed of copper, plastic, or rubber and will range in diameter from ½ to 1 inch, depending on the flow requirements. The tube coils are connected to a boiler, and a pump circulates an antifreeze solution through the coils.

The electric snow-melting system uses insulated resistance wires spaced on a predetermined pattern. The wires are attached to each other with polyvinyl chloride (PVC) bonding cords to form a mat. The mats come in various widths and lengths. The installation contractor can make the mat fit curved areas by cutting the PVC bonding cords. The system can be activated with an on/off switch or an automatic control that monitors temperature and moisture. Depending on the amount of electrical service to the house and the power requirements of the snow-melting system, it may be necessary to upgrade the home's electrical service.

Deteriorating Concrete Driveway

Last May I had a concrete driveway installed. All has gone well with it, but I find its surface deteriorating where the car is parked. It appears that the ice, snow, and salt coming off the car are causing the damage. I sweep the driveway to rid it of slush, and in the process, ½ inch of concrete has been removed.

It appears the driveway was not properly installed. The problem may be that air-entraining agents were not mixed into the concrete. These agents cause microscopically small air pockets to form in the concrete as it cures, and these pockets allow freezing water to expand without damaging the concrete. Concrete used for walks, steps, and driveways should be air-entrained concrete.

Too much water in the concrete mix also could have

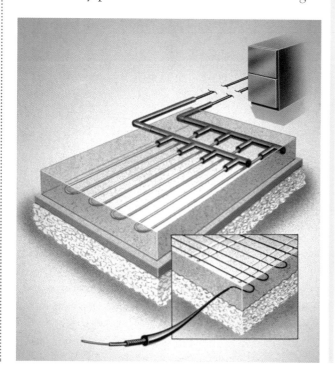

caused the condition; or perhaps the surface was improperly floated (one of the last steps in surface-finishing concrete). In finishing concrete, the water in the mix rises to the surface. The concrete cannot be finished until the water rises to the surface, or it creates a weakened zone below the surface. Also, the concrete cannot be finished with water standing on the surface; the water must be skimmed off or allowed to evaporate.

You should report the damage to the contractor. If the contractor will not replace or repair it, you will have to hire someone or repair it yourself.

Spalling Concrete Driveway

Our six-year-old house has a concrete driveway, the surface of which breaks up during the winter, exposing the rough stone beneath. The contractor says that deicing salt causes this. What can I do to prevent this from happening again?

The condition you refer to is known as spalling. There are three major reasons that concrete spalls:

First, air-entrained concrete might not have been used. Second, the water/cement ratio might have been too high, meaning that the mix was too wet. Third, the concrete surface might not have been finished properly. Water will rise to the surface of the concrete slab as it is finished. This water, known as bleed water, must be allowed to evaporate before final finishing. If the surface is finished with the bleed water present, the water will be worked into the surface, weakening it.

With respect to these three possible problems, it's probably most important that air-entrained concrete be used. This material contains millions of evenly dispersed microscopic air bubbles. The bubbles act as small shock absorbers and prevent spalling when ice forms on and just below the surface of the concrete.

Gravel Driveway

Our gravel driveway has become difficult to drive on during the rainy season. I thought of fixing it myself, but I've been told the job involves more than spreading gravel over the surface. Also, if in the future I decide to pave it with asphalt, what can I do now to make the job easier and less costly?

A gravel driveway is really little more than just a clearing with some gravel or crushed stone thrown over it. Ruts develop if the gravel bed is too thin or the sub-base (earth) is not graded and compacted properly. Also, the gravel will be pushed onto the lawn unless curbing is installed.

A good gravel driveway begins with a well-graded and compacted sub-base to prevent uneven settling. To form a suitable base for asphalt, excavate to a depth of about 4 inches, then cover the area with a 4-inch-thick bed of ¾-inch-diameter crushed stone.

PATIO, WALKWAYS, AND STEPS

Peeling Patio

I have a concrete patio, which I have painted four times,—twice with water-based paint and twice with oil-based paint. After people walk on the patio, I wash it with plain water. The next day, the paint lifts. Also, I thoroughly washed the patio with soap and acid. I even gave it two coats of water-repellent solution. Please tell me what I am doing wrong and what kind of paint will stay on concrete.

Putting the oil-based paint on the concrete was a mistake. For one thing, the oil reacts with the lime in the concrete and forms a soap, which becomes soluble and loses bond. Also, putting a latex paint on top compounded the problem. The oil paint is fairly flexible but the latex is fairly rigid. The differences in expansion and contraction between the two cause the latex to crack.

It would be best to remove the four layers and repaint the patio. You can use a water-emulsifying paint stripper, which can be removed by washing off after it's applied.

Once the paint has been removed, the concrete should be rinsed thoroughly and allowed to dry. Repaint it with two coats of chlorinated rubber-based paint for swimming pools. The first coat should be thinned, so it penetrates and acts as a primer. Follow it with a full-bodied second coat.

You can get the solvents for thinning the paint from your local paint dealer. The solvent will probably contain Xylol or Toluol; both are flammable hydrocarbons, so be careful handling and storing them. A latex paint, which is easier to use, isn't as durable and will wear rather easily compared to the chlorinated paint.

Stair Design

What is the ideal run and rise for porch steps and front-door steps? What dimensions make for a safe and functional stair?

There is no single ideal rise and run. There is, however, an ideal sum of the rise and run. That is, one rise plus one run should add up to 17 to 18 inches. Various state and national building codes give a range for both rise and run, but some don't differentiate between interior and exterior stairs. In order to provide you with an answer that is based on engineering and human-factor analysis, I bypassed the various building codes and went to a book published by the U.S. Department of Housing and Urban Development, "A Design Guide for Home Safety."

The book gives five parameters for interior and exterior stairs. This is based on the fact that exterior stair construction is not always limited by the space restriction placed on interior stairs. Also, exterior stairs are generally used more often by persons carrying bulky or heavy items. The design parameters for both interior

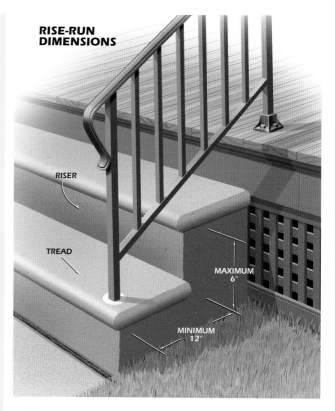

RISE-RUN DIMENSIONS

RISER

TREAD

MAXIMUM 6"

MINIMUM 12"

and exterior stairs for residential applications are listed below. For readers unfamiliar with this terminology, the tread is the horizontal part, from its front edge to its back corner. The run is the tread's dimension minus the width of the "nosing," or the projection at the step's edge.

INTERIOR STAIRS—
- Maximum riser height: $7\frac{1}{2}$ inches
- Minimum run width: 10 inches
- Minimum tread width: $1\frac{1}{4}$ inches
- Maximum nosing: $1\frac{1}{4}$ inches

EXTERIOR STAIRS—
- Maximum riser height: 6 inches
- Minimum run width: 11 inches
- Minimum tread width: 12 inches
- Maximum nosing: 1 inch

A variation in the riser height of steps in a staircase will interrupt the natural rhythm of the person using the steps and could cause him or her to trip and fall. HUD recommends that variation in riser height and tread width in a flight of stairs should not exceed ³/₁₆ inch.

Interlocking Concrete Pavers

I like the look of interlocking concrete pavers, and I would like to install a patio and walkway in my backyard using these materials. Is this a project a homeowner can handle? Where can I get information on installing these pavers?

The use of interlocking concrete pavers is becoming more and more popular around the country. Yes, it is a project a homeowner can handle, but be prepared to invest a lot of sweat in it. There's a lot of digging, shoveling, and bending, and hauling stone, sand, and pavers in a wheelbarrow. But it's worth the effort. The Interlocking Concrete Pavement

Institute (ICPI) says it is not uncommon for these pavements to last at least 30 years.

Like any other pavement, interlocking concrete pavers must be installed above a properly prepared base. The base is a bed of smooth sand over a layer of gravel or crushed stone. The thickness of the stone base depends on the soil. Low-lying, wet soil requires a thick base—perhaps as much as 6 inches. Well-drained areas require only 3 inches. The pavers are installed tightly by hand but without mortar so they can move without cracking in freeze-and-thaw cycles.

Finally, patios and walkways installed next to a house should slope away from the dwelling at a rate of 1 inch for every 8 feet of paved width. This allows rainwater to drain readily from the paved area.

Slippery Concrete Paver

The concrete paving blocks in my yard are covered in algae, and when it rains the blocks are slippery. I had them cleaned with a pressure washer two years ago, but now they are even worse. What can you suggest?

In order to control the growth of algae, you will have to minimize the amount of moisture that gets into the concrete pavers. Coating the pavers with a siloxane-based water-repellent sealer can do this. The sealer should be a breathable type that will not form a film on the surface. Otherwise, the concrete could be damaged from moisture being trapped inside it and then freezing. The surface damage resulting from this is called spalling. Prior to sealing the pavers, you must destroy

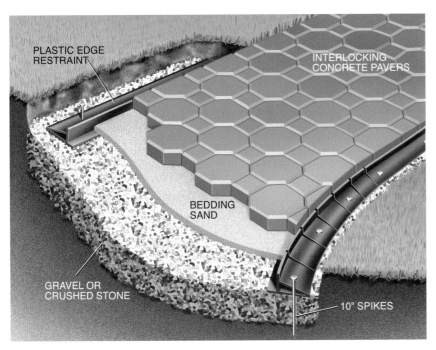

PLASTIC EDGE RESTRAINT

INTERLOCKING CONCRETE PAVERS

BEDDING SAND

GRAVEL OR CRUSHED STONE

10" SPIKES

the algae with a herbicide, but check with the product's manufacturer to be sure it will not damage or discolor concrete.

Slab Jacking

The concrete walk in the corner of our L-shaped house has settled to a slope of 3 or 4 in. The slope causes rainwater to seep into the crawl- space. Can we top this walk with a thin layer of concrete, or will it crumble?

You have three choices: replace the slab, top it with a layer of concrete, or lift it up and fill in underneath it. There are two ways a homeowner can lift a slab: If its edge is accessible, you can use a pry bar. If its edge is not accessible, or it's too big to pry up, you can jack it up.

Span across the concrete slab with two 2 x 4s placed on edge. Bore holes through the 2 x 4s and through the concrete slab as shown (rent a hammer drill if need be). Take some threaded rod and put a spring-loaded wing on the end of each. Push the rod through each hole in the 2 x 4s and into the holes in the slab.

Put a washer and a nut on top of each rod and thread down the nut. Drive down the rod, with a hammer if necessary, until the wings open under the slab. Be sure to put the nuts on the threaded rod before driving down the rod. Driving the rod will mushroom the threads and make it difficult to thread the nuts on the rod.

Tighten the nuts against the 2 x 4s to jack up the slab. Then pour a slurry of cement, sand, and water through the open holes in the slab to fill the void underneath. Turn the rod out of the nuts when the filler under the slab has set up a little. Patch the holes, and the job is done.

Cracked Concrete Walk

Two years ago, I had a porch slab and an adjoining concrete sidewalk replaced. The porch slab was poured on a 4-foot-deep foundation of blocks and concrete. The walk was poured in one continuous piece with the slab, without a space being left between the walk and the slab. Now the walk has cracked badly near the slab after just one winter. Was the sidewalk installed correctly?

Apparently, there is no isolation joint where the slab and sidewalk meet. This would accommodate the slight movement that takes place between them. The slab has a foundation down to the frost line and is not affected by freezing temperatures.

However, the sidewalk is on grade, so any dampness in the base below the sidewalk will freeze and expand, resulting in some frost heaving during the winter. When this movement—or movement caused by settlement or

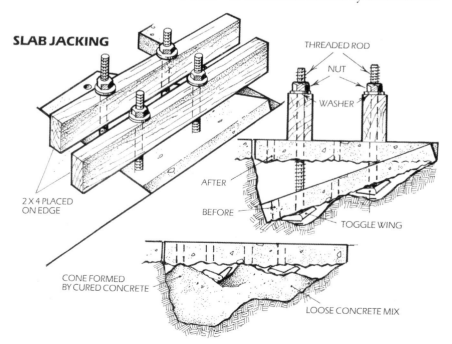

SLAB JACKING

2 X 4 PLACED ON EDGE

CONE FORMED BY CURED CONCRETE

THREADED ROD

NUT

WASHER

AFTER

BEFORE

TOGGLE WING

LOOSE CONCRETE MIX

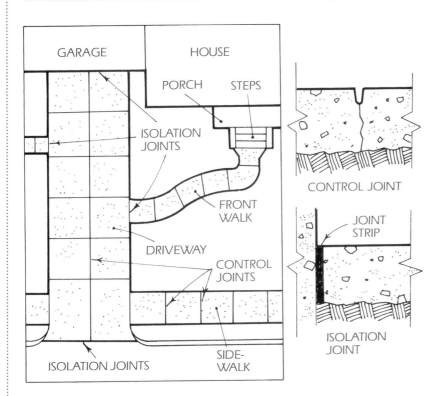

at times the slab is still too slippery. Is there a way to make the slab less slippery?

Two do-it-yourself methods for reducing the slipperiness would be to paint it with a nonslip outdoor paint or to roughen it. The paint, however, may lift or be worn away in areas near the hoop, so roughening the slab may be more attractive. There are two ways to do this: sandblasting and acid-washing. I hesitate to recommend the former, because it is extremely messy—and expensive since it requires the services of a sandblasting contractor.

Acid-washing the surface removes some of the cement paste and exposes more of the sand in the concrete. This results in a grittier texture that is similar to a medium grade of sandpaper. Acid-washing is done with muriatic acid and water. This acid is sold at hardware stores and home centers. Combine 1 part acid with $2\frac{1}{2}$ to 3 parts clean, cool water. Wear old clothing, rubber gloves, and a face shield when working with muriatic acid. Never pour the water into the acid. Instead, add the acid to the water. Use I gallon of the mixture for every 50 to 60 square feet of court area.

Prior to acid-washing, prepare the basketball court surface by washing it with clean water. While the concrete is damp, spread the acid mixture over the area by carefully pouring it from the container. Use a stiff-bristle broom or driveway squeegee to move the mixture over the concrete. The acid solution will bubble and foam as it etches the concrete. Apply more mixture in places where little of this reaction is occurring. The goal is to etch the surface evenly.

After about 20 to 30 minutes, remove the acid-water mixture by thoroughly washing down the area with fresh water. If the surface texture is not as gritty as desired, repeat the process. The process may damage the lawn adjacent to the court. In this case, fertilizing and seeding will be necessary.

shrinkage—is restrained, the concrete will crack.

There are two basic types of joints used in walks, patios, and slabs: isolation joints—often called expansion joints—and control joints—often called contraction joints. The illustration below shows typical locations for each joint.

Isolation joints permit slight vertical and horizontal movement along lines that separate dissimilar concrete emplacements. They are formed by installing strips of asphalt-impregnated composition board.

Control joints create straight-line planes of weakness in the concrete. As a walk or driveway shrinks, a crack forms below the joint in a straight line rather than creating unsightly random, patterns.

Slippery Concrete

I recently built a half-length concrete basketball court in our backyard for my son. I finished the slab with a light broom texture, but

Cleaning Concrete

My house's concrete driveway and walkways have turned black, and besides the black areas, rust spots have formed. I have tried without success to remove the stains with trisodium phosphate and bleach. What cleaner will work?

A fungus causes the spotted black areas, probably mildew, and the rust color stains are iron oxide, which may come from a lawn sprinkler, lawn fertilizer, or iron particles in the concrete.

The cleaners you used are correct for removing mildew, but you may not have used the right proportions. According to the Portland Cement Association (PCA), a solution for removing mildew from concrete is: 1 ounce of laundry detergent, 3 ounces of trisodium phosphate, 1 quart of laundry bleach, and 3 quarts of water. Apply this to the area with a soft brush and let it stand. Then hose the area.

The rust stains can be removed with a solution of 1 pound oxalic acid per gallon of water. Rinse the area off after 3 hours using clean water and by scrubbing with a stiff bristle broom. Oxalic acid, otherwise known as wood bleach, is sold at hardware stores, paint stores, and from some woodworking supply catalogs.

Paint Mist on Concrete

I recently had a cyclone fence painted. Much to my dismay, instead of using a brush, the contractor sprayed the fence and got paint mist on my concrete sidewalk. Is there a way to clean the walk besides sandblasting?

If the mist is slight, then foot traffic and weather will slowly remove it. If the walk has a substantial amount of paint on it, then try applying a paste made with gelled commercial paint remover and talcum powder. Leave the paste in place for 30 minutes, and then gently scrub the area to loosen the paint particles and wash them off with water. It may be necessary to repeat the treatment.

Brick Pavement

My house is 15 years old and well built except for the concrete work. My patio is scaling off in areas. I also have a downstairs garage with an entrance and adjacent areas that look as if they are made of loose gravel. I would like to resurface both the patio and downstairs garage area with brick but do not know how to prepare the surfaces.

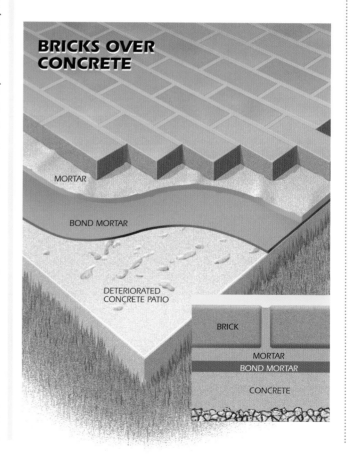

BRICKS OVER CONCRETE

MORTAR

BOND MORTAR

DETERIORATED CONCRETE PATIO

BRICK

MORTAR

BOND MORTAR

CONCRETE

With brick as a paving material there are a number of items to consider in addition to the surface preparation of the base. Brick selection should be based on weather and abrasion resistance. Since most of the paved area will be exposed to the weather, you should select Class SX paving bricks because they are intended for use where the brick may be frozen while saturated with water. This will minimize deterioration due to cracking or spalling.

The concrete base that the bricks are placed on should be at least 4 inches thick. It should be power washed, then its cracks should be filled.

The bricks should be set in Type S mortar—for the ingredients of Type S mortar, visit the Brick Industries Association (BIA) web site. The mortar also helps to level out any irregularities in the base. To improve the bond between the concrete slab and the mortar setting bed, use a bond mortar consisting of portland cement mixed with water or a latex additive. The mixture should have a creamy consistency.

The spaces between the bricks can be filled with mortar, or the spaces between the bricks can be left open temporarily (the bricks are held in place with mortar on their bottom face). The spaces between the bricks are then filled using grout—a thinner version of mortar.

Removing Paint from Brick

I would like to improve the appearance of my brick front steps by removing paint stains from them. If the stains can't be removed, can the steps be painted to give them a more uniform and pleasing appearance?

Paint stains can generally be removed with a gel-type commercial paint remover or with a strong solution of the powderized cleaner trisodium phosphate (TSP) mixed with water. TSP is sold at paint and hardware stores. Mix 2 pounds of TSP in I gallon of water. When using either of these removal materials, wear rubber gloves and eye protection. Apply the remover and let it remain in place long enough to soften the paint, then scrape the area clean. Rinse away any remaining residue to finish the job.

I would not recommend painting brick steps. The paint would be ex-posed to weather extremes and foot traffic, and it's likely that the layer of paint would wear away prema-turely. However, the sides of the steps (and other parts that do not see direct wear from foot traffic) can be coated with acrylic masonry stain. This product is sold at paint stores and home centers.

4. Walls, Windows and Doors

MASONRY WALLS • VINYL SIDING • ALUMINUM SIDING
WOOD SIDING • WINDOWS • DOORS

MASONRY WALLS

Tuckpointing Mortar Joints

My brick home was built in 1963, and the mortar is coming out from between the bricks on the gable end wall. Can the mortar, by itself, be repaired?

It's very important to maintain the integrity of the mortar joints in a brick wall. Otherwise, deteriorated mortar joints become a conduit for water leakage through the wall. There are two repair methods that are considered effective in stopping water migration through a brick wall. One is grouting and the other is tuck-pointing.

Grouting is normally used on hairline cracks in mortar joints. In this method, a thin mixture of portland cement, lime, and sand is applied to the mortar joints using a stiff fiber brush. Two coats are applied, and the grout is forced into the cracks. The problem with this method is that the grout is smeared over the bricks' surface, which drastically changes the appearance of the wall. Tuckpointing is better than this method, because a tuck-pointed wall looks neater and is more durable than one that has been grouted.

Although we can't take the space here to describe tuckpointing completely, here's an overview of the process: The deteriorated mortar is removed to a depth of about $\frac{3}{8}$ inch or to the point where solid mortar is reached. The mortar is removed using a tuckpointing chisel hit by a metal-striking hammer

TUCKPOINTING

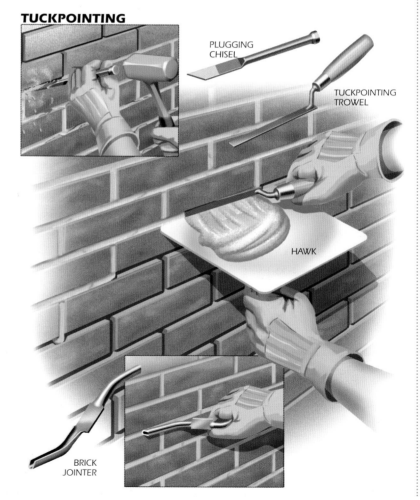

PLUGGING CHISEL

TUCKPOINTING TROWEL

HAWK

BRICK JOINTER

such as a small handheld sledge (not a claw hammer or brick hammer). Debris and dust are brushed from the joints and the joints are gently hosed clean. When surface moisture has evaporated, the joints are packed tightly with tuckpointing mortar in three successive layers and tooled with a brick jointer to match the original profile.

Tuckpointing mortar is made with carefully measured ratios of cement, lime and sand. For a full explanation of tuckpointing, contact the Brick Institute of America.

Chalky Bricks

I'm writing to you about our brick home. We bricked it two years ago. Every now and then, a white chalky substance appears on it. We would like to know if there is something we can do about it.

The white substance is deposits of water-soluble salts, called efflorescence. These salts may be contained in the bricks, the mortar, or possibly the masonry behind the bricks. When efflorescence appears on a brick wall of relatively new construction, it's often the result of water absorbed by the bricks. The soluble salts dissolve in water (from rain, snow, or ice) and are deposited on the brick surface when the water evaporates.

However, it's also possible that water is entering the wall through open joints or poor construction methods and has found a path to the surface.

When there is a sufficient amount of water to dissolve the salts, there will be an efflorescence buildup on the wall surface. In this case, you should examine the wall and construction details for open joints and possible sources of moisture penetration. All open joints must be sealed and poor construction methods corrected.

Efflorescing salts will generally disappear of their own accord with normal weathering, especially when they are the result of absorbed water. However, you can remove the salts by dry-scrubbing with a stiff brush. Heavy accumulations or stubborn deposits may

be removed by scrubbing with a solution of one part muriatic acid to 12 parts water. It is imperative that the wall be saturated with water before and after the solution is applied.

Faded Black Mortar Joints

My concern involves the black mortar used on the brickwork of my 10-year old home. This mortar was used to accent the bricks we used. When it was new, the mortar was very black. However, after the first year it turned gray. Now, after 10 years, the sun and weather have turned the mortar almost natural. Is there anything I can do to restore the black? In every home I've seen with similar mortar, it has become faded. Please let me know if there's anything I can do about this.

Although nothing can be done to restore the black in the original mortar, there is something you can do to produce black mortar joints. You can either stain the joints or tuckpoint them. Both of these methods are tedious and labor intensive, so unless you do the work yourself it can be quite an expensive undertaking.

According to the people at the Portland Cement Association, the original mortar mix probably had carbon black to produce the black color rather than a mineral oxide. They say that carbon black is sensitive to ultraviolet light, not stable, and has a life of about seven months to a year.

If you want to tuckpoint, you will have to grind out the existing joints to a depth of about $5/8$ inch and then fill them with a new mortar containing a black mineral oxide pigment.

The other method for producing black joints is to stain them using a water-repellent, penetrating masonry stain. The stain can be applied using a $1/2$-inch brush. You must work carefully, however, so as not to smear stain on the bricks.

Caring for a Fieldstone House

We just purchased a home built in 1935. It was constructed by a man from Finland and it has a fieldstone exterior. I am concerned about how to take care of the stone. Someone told me that I should spray the exterior with silicone. The home has been well maintained and there are no particular problems with the stone and concrete to date. I would appreciate any advice that you may have to give.

The problem with fieldstone exterior walls is the mortar joints between the stones and not the stones themselves. The way to care for these walls is to periodically inspect the condition of the mortar joints between the stones. Spraying the wall with silicone will not help defective joints, as the spray will not bridge a crack or open joint. Openings in the joints will permit water penetration during a rain. In the winter months, the water trapped in those areas will freeze and expand, causing further damage.

When inspecting the joints, look for cracks and sections where the mortar is not bonded to the stones. If you find any openings, they should be repaired with a mortar mix that is compatible with the surrounding mortar. This is probably best left to a mason familiar with masonry repair on historic buildings.

Cracking Brick Corners

My 12-year-old brick home is cracking at the corners, and some of the bricks are loose. The contractor who looked at it for me is perplexed. Do you have any ideas?

The fact that the cracking has occurred at the corners of the house is significant. The gutter downspouts, which direct the roof rain-runoff, are usually located at the corners of the house. In a proper installation, each downspout is either connected to underground piping to carry

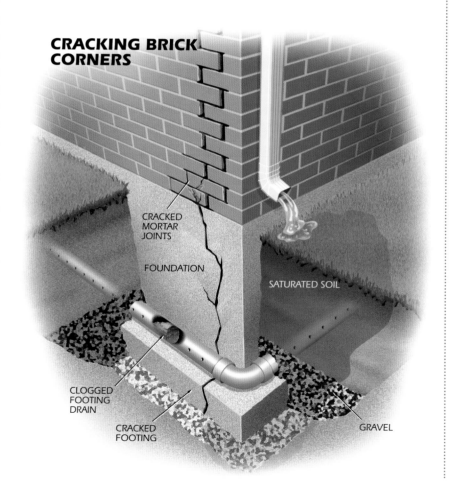

CRACKING BRICK CORNERS

CRACKED MORTAR JOINTS

FOUNDATION

SATURATED SOIL

CLOGGED FOOTING DRAIN

CRACKED FOOTING

GRAVEL

the effluent away from the house or fitted with an elbow and splash plate to direct the discharging water away.

If the elbow has fallen off or is missing, the water from the gutter will discharge directly below the downspout and will accumulate around the corner of the foundation. Over a period of time, this could undermine the foundation footing in the corner, which will cause uneven settlement. This in turn will cause cracking in the foundation and, subsequently, the bricks and mortar.

If there is a functioning footing drain around the foundation, this water will be carried away. However, some footing drains are damaged during construction and some become clogged over time. Once you direct the downspout effluent away from the house, you should resolve the cracking problem. Loose bricks should then be reset and cracked joints repointed.

Spalled Bricks

The exterior brick of our home is porous and is starting to crumble. We have had it tuck-pointed and sprayed with a transparent water repellent. This has helped very little. Would painting the bricks with a vinyl or lucite paint stop the crumbling? Our home is 26 years old and otherwise in very good condition. Can you please help us?

Unfortunately, the only solution is to replace the deteriorated bricks. Water penetrates through cracks in the mortar joints. The water freezes and thaws, causing the bricks to spall.

Once the bricks spall, you cannot reverse the condition. You can only reduce further spalling by sealing cracked or open mortar joints.

Sealing the brick face with a clear, penetrating sealant is not recommended by the Brick Institute of America. It can cause more harm than good by trapping moisture that was present in the brick. This moisture freezes and causes the bricks to spall.

Painting the bricks is a stopgap measure. The paint can act as an adhesive, keeping the crumbled pieces together. When the adhesive properties of the paint reach their life cycle and the paint peels off, the crumbled pieces will peel off with it.

Removing Ivy from Brick

My wife planted some ivy on an outside brick wall. The ivy has since died and left dried, woody vines clinging to the brick. How can I remove them without damaging the brick?

The Brick Industry of America trade association recommends cutting the vines off the wall with shears, then removing the suckers that hold them to the wall. To remove the suckers, dampen the wall with a garden hose and then use some laundry detergent and a stiff fiber brush to scrub the suckers off. Rinse the wall thoroughly.

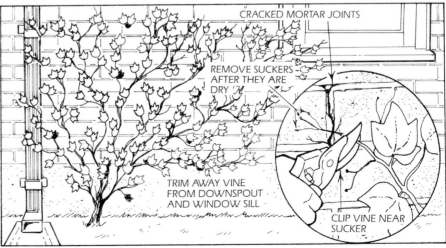

CRACKED MORTAR JOINTS

REMOVE SUCKERS AFTER THEY ARE DRY

TRIM AWAY VINE FROM DOWNSPOUT AND WINDOW SILL

CLIP VINE NEAR SUCKER

MASONRY WALLS

The longer the suckers and vine are left on the wall, the more difficult the vegetation is to remove. It can oxidize and harden to the point that removing it will damage the wall. Even if that is not the case, you should thoroughly inspect the wall for damage that the vines may have done to the mortar joints. The joints may require tuckpointing to repair them.

I recently purchased a 27-year-old brick house that has one side covered with vines. I just don't know whether I should let them grow or remove them so they don't damage the wall. Do you have any recommendations?

Many people feel that vines enhance the beauty of a building and they don't care about the problems that they can create. Vines, although aesthetically pleasing, are undesirable because they can conceal termite shelter tubes, nesting insects and cracked mortar joints. They can widen cracks, loosen shingles, and crush downspouts. Also, the dampness associated with vines promotes rot in framing and wood trim.

If you decide not to remove the vines, keep them trimmed away from the roof edge, gutters, downspouts, windows, and wood trim.

The Brick Institute of America recommends carefully cutting vines away from the brick rather than pulling them off. Pulling off vines can damage the brick and mortar joints. The suckers that attach and hold the vines to the wall will be left on the wall. Wait about two or three weeks until the suckers dry up and turn dark. Then remove them with a stiff brush and some laundry detergent. If you wait too long before removing the suckers, they may rot, oxidize, and become so hard that removing them will damage the wall.

Ivy Footprints

My wife loves the appearance of the English ivy growing on the side of our red brick home. But I have noticed while maintaining these vines and cutting them back that they leave

"footprints" on the brick and mortar. I have tried removing these marks with a wire brush, but this is ineffective. Is there a way to remove them without damaging the brick and mortar?

You may not be able to remove the footprints without damaging the wall. It depends on how long they've been on the wall since you pulled the vine away. The Brick Institute of America recommends that you wait two to three weeks after cutting the vines for the vine's suckers to dry up and turn dark. Then remove them with a stiff bristle brush and a solution of water and laundry detergent. If you wait too long before removing the suckers, they may rot, oxidize, and become so hard that removing them will damage the brick and mortar.

Cleaning Bricks

We moved into a lovely brick home seven years ago, but we have a problem with its bricks. Lower sections of the rust-colored bricks have a white deposit on them. This apparently is caused by water from the lawn sprinklers splashing on them. What can we do to renew the look of the bricks?

The white deposits are water-soluble salts known as efflorescence. The salts usually exist within the bricks and possibly the mortar joints. When water is absorbed by the brick wall, the soluble salts enter into solution, and efflorescence is formed as evaporation takes place from the surface of the bricks.

These salts are relatively easy to remove, compared with stains caused by iron, smoke, oil, or tar. In most cases, the efflorescence can be removed by brushing with a stiff brush and water. However, if there are stubborn deposits of salts or a heavy accumulation, removal may require scrubbing with a solution of muriatic acid. Use 1 part acid diluted in 12 parts of water. It's important to saturate the wall before and after the solution is applied.

VINYL SIDING

Mold on Vinyl Siding

Some of the vinyl siding on my house has a green mold growing on it. What causes it, and how do I get rid of it?

It's unusual to have a green mold on vinyl siding, although some molds are green. The most common form of mold is mildew, which is black. I suspect that the growth on the siding is moss, in which case you simply need to wash the siding to clean it.

Mix the following solution: $\frac{1}{3}$ cup powdered laundry detergent, $\frac{1}{3}$ cup powdered household cleaner, $\frac{1}{3}$ gallon water and 1 quart of liquid laundry bleach.

To apply, use a long-handled car-washing brush or the type of long-handled brush that attaches to a garden hose. To avoid leaving streaks, wash the siding from the bottom to the top and then rinse the siding to complete the job. Some areas may require multiple cleanings.

Stained Vinyl Siding

Last spring, I sprayed a swarm of wasps with wasp-and-hornet spray. The spray splattered on my house's vinyl siding and badly stained it. I have tried everything I can think of to remove the stain, to no avail. Do you have any suggestions?

There are a few things that can permanently stain vinyl siding. Unfortunately, petroleum-based insect spray is one of them. The spray's petrochemicals react with vinyl, discoloring it. I spoke with a manufacturer of vinyl siding, who said you might be able to lighten the stain by cleaning the area with a soft cloth dampened with naphtha or mineral spirits and rinsing it with water. But he cautioned that a sheen may be left on the area afterward.

Depending on the location of the discoloration, you might consider hiding the area with shrubbery. Another option would be to replace the section of siding.

REMOVING STAINS FROM VINYL SIDING

STAIN	CLEANERS
BUBBLE GUM	Fantastic, Murphy Oil Soap, Windex, solution of 30% vinegar and 70% water
CRAYON	Lestoil
OIL-BASED CAULK	Fantastic
FELT-TIP PEN	Fantastic, water-based cleaners
GRASS	Fantastic, Lysol, Murphy Oil Soap, Windex
LIPSTICK	Fantastic, Murphy Oil Soap
LITHIUM GREASE	Fantastic, Lestoil, Murphy Oil Soap, Windex
MOLD AND MILDEW	Fantastic, Windex, solution of 30% vinegar and 70% water
MOTOR OIL	Fantastic, Lysol, Murphy Oil Soap, Windex
OIL	Soft Scrub
PAINT	Brillo Pad, Soft Scrub
PENCIL	Soft Scrub
RUST	Fantastic, Murphy Oil Soap, Windex
TAR	Soft Scrub
TOPSOIL	Fantastic, Lestoil, Murphy Oil Soap

Fortunately, other stains can be readily removed with common household cleaners. The table of stains on the previous page was prepared by the Vinyl Siding Institute, and it's based on a study the organization conducted of household cleaners and the effect the cleaners had on the siding's appearance after two years of outdoor exposure.

Cleaning Vinyl Siding

The vinyl siding on our garage walls has mildew on it that is very difficult to remove. I've tried a specialty cleaner, and I've also used chlorine bleach. Neither worked. What can I do?

According to the Vinyl Siding Institute (VSI), you can clean mold and mildew from vinyl siding using Fantastic, Windex, or a solution of 30 percent vinegar and 70 percent water. The institute does not recommend using aggressive or highly abrasive cleaners because they can damage the siding's surface.

Advantages of Vinyl Siding

Rather than painting our house's wood siding, we are considering having vinyl siding installed over it. Can you tell us about the advantages and disadvantages of vinyl siding?

The problem with vinyl siding is not the siding itself but the installation of the siding by nonqualified contractors. Vinyl siding expands and contracts as the temperature changes. When the siding is improperly nailed, this movement usually results in waviness and blisters in the vinyl panels.

But, like any building material, vinyl siding does have its advantages and disadvantages. Vinyl siding normally does not dent from impact—it merely flexes and springs back to its original shape. However, during cold weather the siding becomes brittle, and a hard blow could crack it.

Another advantage to vinyl siding is that its coloring is the same throughout its thickness. If the siding is scratched, it is less visible than on aluminum siding or wood siding. With aluminum siding, the scratch would reveal the shiny aluminum underneath, and with painted wood siding, the wood itself is visible beneath a scratch.

Venting Vinyl Siding

Two years ago, I installed vinyl siding over the 22-year-old hardboard siding on my home. Now I have a moisture problem. In the winter, icicles form around outdoor light fixtures. What's the best way to vent this siding?

It should not be necessary to install vents in the vinyl siding itself. If the siding is applied correctly, you should not have a condensation problem. The siding allows moisture to escape through prepunched weep holes along its bottom edge. Check with the company that installed the siding. It's possible that the weep holes have been covered.

The problem may not be with the siding. Instead, the problem you have in the winter may be caused by an ice dam. This occurs when heat lost through the attic melts snow on the roof above. A trickle of water runs under the layer of snow and freezes at the roof's edge, which is colder than the main part of the roof. Often, ice formed in this manner extends under the soffit and down the siding. Ice dams can be prevented by insulating and ventilating the attic and sealing holes through which heat enters the attic from the living space.

Vinyl Siding over Asbestos

We live in an old Victorian house with asbestos siding on it. We want to put vinyl siding on it, but we need to know if it is necessary to remove the asbestos siding first. Siding contractors say they can apply the vinyl over the asbestos, but one contractor warned us that the asbestos siding will crack when the vinyl siding is nailed on top of it, and that pieces of broken asbestos would fall down and accumulate, causing bulges in some areas of the vinyl siding.

I discussed your situation with the Vinyl Siding Institute, which said that it recommends not disturbing the asbestos shingles. The best practice would be to apply sheathing over the existing asbestos siding and then apply vinyl siding over the top of the sheathing. The sheathing will serve to flatten the walls, and will contain any broken asbestos shingles. The vinyl siding should be applied using nails that are long enough to penetrate the sheathing, the asbestos shingles, the sheathing below the shingles, and the wall studs.

Vinyl and Asbestos Siding

We own a 1950s-era house that has asbestos shingles on the sides and back and vinyl siding on the front. The shingles are in good condition, but we are concerned that they will be a factor in selling our home. If the shingles are removed, will their disposal be difficult?

Although the shingles are referred to as asbestos, they are really asbestos cement shingles. That is, the asbestos fibers are encapsulated in cement. As long as the shingles are in good condition, there is no problem. However, if they are abraded or sawn, asbestos fibers can be released into the air.

If the shingles are in good condition, they are generally not a factor when the home is sold. Nevertheless, there will always be buyers who try to make the shingles a part of the sales negotiation. Even if the shingles are in poor condition, they don't have to be removed. You can install new siding over them.

Removing the shingles is expensive, since the work must be done by a licensed asbestos-removal contractor. The waste will have to be disposed of in a landfill designed to take this material.

ALUMINUM SIDING

Aluminum versus Vinyl Siding

We want to install new siding on our home. We are wondering which siding you would recommend, vinyl or aluminum?

I don't have a preference. No siding material is perfect, and your choice should be based on which siding, in your opinion, has more advantages than disadvantages.

Aluminum siding comes smooth or embossed with wood grain. It is relatively maintenance-free, corrosionproof and termiteproof, and it won't rot. Its surface finish is a durable, baked enamel paint.

If aluminum siding is scratched, the silver color of the bare aluminum below is exposed. However, the scratch can easily be repaired with touchup paint. One problem with aluminum siding is that it can be dented, say, by a baseball striking it or a ladder pressing on it.

Vinyl siding is very much like aluminum siding in size, shape, application, and appearance. Vinyl siding normally does not dent from impact. However, during very cold weather, the siding becomes brittle, and a hard blow could crack or shatter it. Vinyl siding expands and contracts as the temperature changes. If they are improperly nailed, this movement can result in waviness in the vinyl panels.

Loose Siding Repair

A section of the horizontal-lap aluminum siding has loosened around the waist of our Two-story, 10-year-old home. I've been told that the siding will have to be taken down and reinstalled. I can't believe there isn't some way to remedy this without going through that kind of work and expense.

I checked with the technical department of the trade association of aluminum-siding manufacturers, the American Architectural Manufacturer's Association (AAMA). The Association recommended the following repair procedure:

1. Cut the panel that's above and adjacent to the loose panel along its entire length at a point just above its center. This can be done using a utility knife and tin snips. Then remove and discard the lower section.
2. The nailed portion of the loose panel will now be exposed. Remove the loose, bent panel and replace it with a new panel. Make sure to nail it properly at the top and then to snap the bottom section into the top lock of the siding panel below.
3. Remove the top lock of a new panel by deeply scoring it with a utility knife. Bend the piece and snap it off.
4. Apply a heavy coat of gutter seal along the full length of the cut panel.
5. Install the new panel over the gutter seal. Tuck the top of the panel under the lock and snap the bottom lock in place. Be sure the gutter seal makes contact with the new panel. Apply pressure with your palm, but be careful not to bend the panel.

Painting Aluminum Siding

I have a question about painting the aluminum siding on my house. It's in good shape, but it's a really ugly color. What kind of paint should I use? Also, do I need to use a special primer and tools?

Aluminum siding can be painted with a quality exterior latex paint. But there are several things you need to consider before painting. First, you will need a primer only if the siding's original coating is in poor condition. Choose a primer that is intended to work with the topcoat that you are using. It sounds like you want to drastically change the color. In that case, buy enough paint to apply two coats. The paint store can advise you on quantity.

Before applying the paint, you must remove all chalked paint, dirt, and mildew from the siding. This is best accomplished by combining a cleaner with power washing and scrubbing. You can do this yourself with a rental machine, or have it done professionally.

Aluminum Siding Repair

How can I repair the bottom two courses of aluminum siding on my home? I don't want to remove the siding all the way down from the top. I have several courses of new, unused siding in storage that I can use for the repairs.

It isn't necessary to remove the aluminum siding from the roof level down, in order to repair the two damaged sections of siding at the bottom of the wall. The following procedure is based on recommendations by the American Architectural Manufacturer's Association (AAMA), the trade group of aluminum building products manufacturers.

Cut the top damaged panel along its length at a point just above its center. This can be done using a utility knife and tin snips. Remove and discard the lower section of this panel. The nailed portion of the bottom damaged panel will now be exposed. Remove the panel and replace it with one of your new panels. Make sure you nail it properly so that it can expand and contract with temperature changes.

Apply a heavy coat of gutter seal along the full length of the cut panel. Remove the top lock portion of the new panel by scoring it with a utility knife and snapping it off.

Install the new panel over the gutter seal. Tuck the top of the new panel under the lock of the panel above and snap the bottom lock in place. Be sure the gutter seal makes contact with the new panel. Apply gentle pressure with your palm. Also, do not nail the repair panel in place.

Aluminum Corrosion

I have brown, aluminum vented soffits on my house that were installed when the house was built. Very soon afterward, the cut edges of the vent slots started to form a white powder. Now, on one part of the house, the soffit is almost completely white. Is there something available to remove this unsightly corrosion, or do I have to replace the soffit? I've seen much newer houses with this problem.

I contacted several aluminum manufacturers and associations about this. They all said that it was very unusual, and none of them had a solution. Aluminum corrosion is common in salt atmospheres, such as along coastal areas and in areas with acid rain.

If the white deposits are minerals, they can be removed with a commercial aluminum cleaner. If the deposits are caused by corrosion, then the only remedy is to replace the soffits with ones made from a different material.

Painting Galvanized Metal

I need advice on how to paint galvanized-metal siding. The siding was last painted 25 years ago. What would be the best type of paint to use, latex or oil? Or should I try an automotive enamel? The building is a garage, and I would like its color to match other buildings on the property.

I'd recommend using latex paint. It's easy to work with, and the brushes and rollers can be cleaned in soap and water. Use a primer that is formulated for galvanized metal. Since the metal siding has been exposed to dirt and dust, it's important to clean it thoroughly and follow other surface preparation instructions that are listed on the can label. The topcoat over the primer can be either latex or oil. Choose the type that can be tinted closest to the color of the other buildings. Also, it's always advisable to use a primer and topcoat from the same manufacturer.

WOOD SIDING

Dry, Cracked Board Siding

Two years ago, I bought an 18-year-old house with board and batten siding (vertical 1- X 10-inch boards with 1 x 3-inch strips covering the joints). I think the wood is cedar. It is very dry and has cracked and warped in many places. When I nail the wood back down, it only cracks more. I plan to replace the 1 X 3 strips and hope this will hold the warped boards tight. What kind of finish should I use?

Based on your description, it sounds as if any attempt to correct the problems would be fruitless. Pulling the boards down using new 1 X 3 batten strips may straighten them temporarily, but it will also crack them. Warping of the siding is probably a result of inadequate nailing. With vertical siding, backer blocks are normally installed between the studs to provide a good nailing base. With a poor nailing base, when one side of the board is more moist than the other, it expands, pulls the nails loose and warps.

I don't think the siding is cedar. The warping and cracking that you describe is not typical of cedar. It may be spruce or hemlock. In any event, the badly distorted boards and battens should be replaced and fastened into a good nailing base. If more than one-third of the siding needs work, you would be better off re-siding completely. If you only replace sections, you will be unable to match the weathered look of the original

boards. To minimize this, coat all walls with a semi-transparent, penetrating oil-base stain

Wood Shingle Treatment

Our 15-year-old house has cedar shingle siding that has weathered to a pleasant gray. The shingles have never been treated with a preservative, oiled, or stained. Some have curled, a few badly. We like the way the house looks, and we want to improve the shingles' durability while maintaining their pleasant gray color. We also wonder if weathered shingles present a fire hazard.

First, wood-shingle siding—even that which is very weathered—is not considered to be a fire hazard. Of course, shingles will burn in the event of a fire, just as any wood siding and trim will.

On the other hand, although a wood shingle roof is not a fire hazard, it is vulnerable to catching fire from embers thrown out the top of a fireplace chimney. Treating a wood-shingle roof with fire-retardant chemicals increases its fire resistance.

Cedar shingles are very resistant to decay. Good-quality shingles on exterior walls will last for at least 50 years. Preservative treatments are required only under extreme weather conditions. Since you like the looks of your house, there is nothing you have to do with shingles that are in good condition. However, badly curled shingles should be replaced and the replacement shingles should be treated to blend into the surrounding shingles.

New replacement shingles can be made to look weathered by coating them with bleaching stains that are sold at hardware stores and paint stores. You can also make a bleaching agent that was developed years ago by the Texas Forest Products Laboratory. Mix one pound of baking soda into a gallon of warm water. Dip the shingles into the solution and place them in sunlight for four to five hours. Experiment with a few spare shingles first, and dilute or concentrate the solution if necessary. When the shingles are dry, wash off any powdered residue and install them to match the pattern of the surrounding shingles.

To Stain or Not to Stain

I have a question that needs an unbiased answer. I have a contemporary home that I built seven years ago. The exterior is rough-sawn cedar. It was stained with a wood-preserving transparent stain because I wanted an "instant" weathered look. This is a one-story house with a hip roof and a 3½-foot over-hang all the way around. With the foundation shrubbery, not much weather gets to the siding.

My question: Does it need restaining? Painters and paint store owners tell me it needs restaining. I don't think it needs it, but I want to do what's best for my home. Your unbiased professional opinion would be greatly appreciated.

The answer depends on the extent to which the siding is exposed to the weather. All species of wood that are exposed to the weather swell and shrink, depending on moisture gain or loss. This, in turn, causes the surface to crack and check. When wood is exposed to the sun's rays over a period of time, the cellular structure is affected to the extent that naturally weathered cedar wears away at a rate of ¼ to ½ inch of surface per 100 years. You can retard natural weathering by coating the cedar with a good-quality stain containing a water-repellent preservative. However, if your siding is indeed protected from the weather as you describe, then coating it should not be necessary unless you want to do it for aesthetic reasons.

The main cause of problems with all types of wood siding is constant moisture. As long as the wood is kept

dry, it can last indefinitely. However, when the moisture content exceeds 20 percent, the wood will rot. The area of concern with wood siding is the portion near the ground. Although it is not good construction practice, I quite often find siding is either very close to the ground or in direct contact with it. This usually results in rotting sections. Although cedar has a natural resistance to rot, it is not immune to it.

In order to retard the growth of the fungi that cause rot, wood that is exposed to constant moisture should be coated with a good paraffin-based, water-repellent preservative.

Protecting Cedar

What is the best product to protect cedar siding, especially when it's exposed to severe weather?

Cedar siding should be coated with a finish that provides water repellency and protection from ultraviolet light. The finish should also contain a preservative that kills mold and mildew. Ideally, the finish should also allow the wood's grain to show through. For this reason, semitransparent oil-based stain is a good choice. It is lightly pigmented and provides water repellency. The pigments in the stain provide a small amount of protection from the damaging effects of ultraviolet light. You can purchase this stain at a paint store or home center.

Clean the siding and let it dry before applying the stain. Surface cleaning may be done with low pressure power-washing, using bleach in the wash water to kill mold.

Lead Paint Removal

My old farmhouse is covered with clapboard that was painted with a lead-based paint. Can you tell me the most economical and efficient way to remove it? Also, what precautions should I take when doing this work?

There is no economical or efficient way to remove lead-based paint. Repainting with an oil-based paint will not solve the problem because the lead will eventually leach through the topcoat. There are currently three acceptable methods of abating lead paint:

1. Replacement
2. Removal
3. Encapsulating or covering

Replacement is most appropriate for windows, sills, woodwork and doors, but not for exterior siding. Removal of the lead-based paint, which includes scraping the surfaces using hand scrapers, chemical solvents, or heat guns, is the most costly because it is labor intensive and generates large amounts of lead dust, which must then be disposed of in special landfills.

The least costly approach is encapsulating and covering. The exterior walls may be covered with aluminum or vinyl siding. However, the walls should first be treated with an encapsulant to prevent lead-based paint contamination during the siding installation. The encapsulant is sprayed or rolled on. It bonds chemically with the lead in the paint below to prevent it from leaching.

Since you refer to the building as an old farmhouse, I assume that some of the paint is peeling and flaking. If this has been the case, the ground around the building is contaminated from lead dust and flakes. If the soil is contaminated, it should be covered with sod or gravel. It's likely that the walls and trim inside the house have been covered with lead-based paint, too.

Before any work is undertaken, you should engage the services of a trained lead-paint inspector to determine the extent of the problem.

Latex over Oil

Please advise me on the problem with a building that has two coats of oil-based paint and needs painting. I would like to change to a water-based paint. Would this idea be possible? I plan to perform the work myself.

I spoke with some people at a paint company's technical service department, and they said you can spread latex over oil-based paint. In fact, you can go back and forth between latex and oil. However, the wall surfaces must be properly prepared before painting.

The surfaces must be clean and free of dirt and grease spots. The paint should not be excessively chalked. A good way to ensure that the surface is clean is to have the walls power washed. Also, the wash water usually contains additives to kill mildew.

It's not necessary to prime all painted surfaces. Spot prime those areas that have been scraped to remove peeling and blistered paint.

Next, you can apply one coat of a good quality house paint, although two coats are better. A good quality paint, applied correctly, should last six to seven years.

WINDOWS

Insulated Glass Windows

I would be indebted to you for life if you could solve my problem. We had a wonderful ocean view until our double-pane glass sliding doors clouded up. Is there any way to eliminate this awful haze?

We frequently receive this question. Unfortunately, the problem is caused by a faulty seal between the two panes of glass. It cannot be corrected by the homeowner.

To solve this, you must replace the insulated glass. The frame of the door or window, however, normally does not have to be replaced. Depending on how old the window is, you may be able to get a replacement from the manufacturer. Most manufacturers guarantee their products for five to ten years against failure caused by a faulty seal. They will replace a pane at no cost within the warranty's time limit. Some manufacturers have longer warranties.

The construction in a cross-section of a typical insulated glass window. The window consists of two panes of glass separated by a metal spacer usually made of aluminum. The spacer contains desiccants that absorb moisture in the airspace.

The spacer corners are fused or bent to be air-and gas-tight. The glass panes and the spacers are held in position by at least two seals.

Finally, the airspace between the panes is filled with argon or air at atmospheric pressure. It is not under a vacuum. Argon, which is inert and chemically inactive, is used more often than air because it has a lower thermal conductivity. This means it has a greater resistance to heat flow.

Failed Window Seals

How can I repair the failed seal between the insulated glass panes on my windows? I have replaced almost all the windows, and I have replaced some twice. I have repaired some. Some work. Some don't.

Over the years, we've received many letters about failed, insulated glass windows. Repairing them is not a do-it-yourself job, and they should be replaced. Repairing these windows may work temporarily, but the repair usually doesn't last.

If you've replaced almost all the windows—some of them twice—it's obvious that they were not good-quality windows. When properly made, insulated glass windows will last many years. Because windows are an important part of your home, they should never be purchased strictly on price. It's better to buy windows that are made by a manufacturer that has been in business long enough to justify confidence in its product and that backs its product with a substantial warranty.

Foggy Thermal Panes

My daughter's home in Michigan has double-glazed windows. Almost all of them have developed leaks in the vacuum seals and are clouded between the panes. Somewhere I read this fog could be eliminated or alleviated with the injection of a gas. Any help you can give me will be appreciated.

I know of no method for correcting clouded double-glazed windows short of taking them apart, cleaning the panes, and resealing the edges. Injecting a dry gas or drawing a vacuum (both extremely difficult procedures) in the space between the panes may temporarily minimize the condition. But unless the seal failure is located and corrected, the condition will recur. Usually, the failure is just a tiny hole or crack in the perimeter joint, and difficult to find. Often, when there is repeated condensation and drying, a residue forms on the inner surfaces that is almost impossible to clean.

Most insulated panes are made with a hollow spacer around the edge that has an open end facing the space between the panes. The spacer is filled with a desiccant to absorb residual moisture. When there is a seal failure, the desiccant becomes saturated, and moisture condenses on the panes.

To repair a faulty window you would have to remove the existing seal, clean out the edge joints so that a new seal can be applied, replace the dessicant in the spacer, clean the panes, and reassemble the window unit. It's really not a do-it-yourself project.

Your best bet would be to contact the manufacturer of the windows. Most of them have a warranty—usually from five to ten years—against product failure caused by a faulty seal. They will replace a pane at no cost within the first five years, and at a prorated cost for the next five years. Some manufacturers have longer warranties.

Cracked Insulated Pane

In 1990, we bought a new home with double-pane windows. I accidentally cracked the inner pane on one of the windows. This allowed moisture into the pane, causing fogging. I need to know how to repair it. Two separate glass companies said it will cost several hundred dollars, but I don't buy that. Can you help?

Unfortunately, there is no way to fix the pane. It must be replaced. It's not as simple as removing one of the panes, replacing it with a new pane and then sealing it with a caulk gun. Double- and triple-glazed windows have an airtight seal between the panes. This is achieved under carefully controlled conditions at a factory, where the panes are cleaned and sealed together. It's unlikely that a homeowner can duplicate this.

An End to Broken Window-sash Cords

I have double-hung windows in my home, with cords and weights, and I'm getting tired of fixing broken cords. Could you tell me how to replace them with springs? What types of springs are available? Also, after I replace the cords and weights, how do I insulate the hollow space and what type of insulation should I use?

If you like the lifting action of the counterweight on the double-hung window, you should replace the broken sash cord with metal chains, which are available at hardware stores. The chains won't fray or become brittle with age and break.

You can also replace the weights and cords with metal pressure channels. These channels do not provide lifting action, but they have spring-action strips on both side jambs that hold each sash securely in any position. This ensures a snug fit that prevents rattling and minimizes cold-air leaks.

The first step is to remove the inside trim or stop molding from the sides of the jambs and remove the sashes from the frame. Then, cut off the sash cords, remove the old pulleys, and clear any obstructions out of the old channels.

Fit the channels on each side of the sash. Pick up

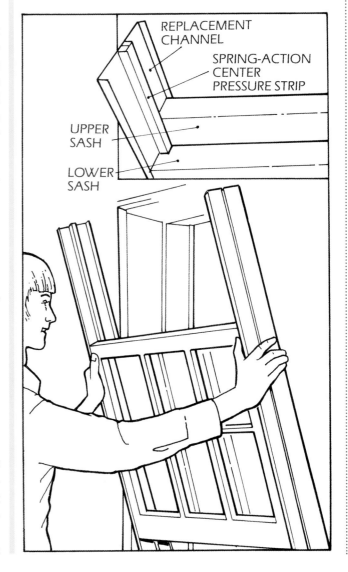

the sash and channels together and place them into the frame. Finally, tack the channels in place and reinstall the trim.

Metal pressure channels are available in hardware stores and home centers. Before installing the channels, the wall cavity for the cord and weights can be insulated by loosely stuffing it with fiberglass or Rockwool. You can also pour a loose fill insulation such as vermiculite into the hollows. If the area is inaccessible, you can fill it using an aerosol foam insulation.

Painting Aluminum Storm-window Frames

The aluminum frames on my storm windows were originally white. However, much of the paint has flaked off. I want to paint them another color. I know I must get all the loose paint off first, but I am unsure of how I should treat them prior to painting, and what type of basecoat and topcoat to use.

The frames must first be cleaned of all dust and dirt. Also, as you mentioned, all loose and flaking paint must be removed. The frames should then be coated with a 100-percent-acrylic latex primer. This should be followed by a good-quality latex topcoat from the same manufacturer that made the primer, because this ensures maximum compatibility between the primer and topcoat.

Before applying the primer you might want to consider sanding down the edges of the original paint to a feather edge at those areas where sections of the paint have flaked off the aluminum frame. If you don't do this, the frames will look good from a distance but on closer examination you'll be able to tell that the surface is uneven because of the original paint. Feathering the edges will reduce the uneven appearance. Make sure to carefully remove all traces of dust left from sanding

before applying the primer. Dust on the surface will prevent proper adhesion of the primer.

Leaking Windows

I have a two-story home with a dormer located above my downstairs bedroom. Water leaks through the bedroom ceiling and runs down the wall. No rain comes in upstairs. I put flashing on the outside wall all around the dormer and then applied tar to seal the joints. It still leaks. I had a roofer look at it, but he had no answer. Any suggestions would be great.

Finding the source of a leak is a trial-and-error process. Since you have eliminated the joints between the dormer and the roof as a possible source of water entry, I suggest you check the dormer window. The window may not have been installed properly. There may be open joints between the window frame and the exterior wall, or possibly open joints in the sash tracks or sill.

You can check for leakage by using a garden hose and spraying all of the joints around the window from a distance of about 2 to 3 feet. Spray the window area and joints for several minutes.

A good way to tell if there is window leakage, and to minimize any further damage to the ceiling and wall in your bedroom below, is to cut a 6-inch-square hole in the drywall just below the left and right corners of the window. If there is insulation (and there should be), move it aside so that the wall sheathing is visible.

Have someone check those areas for dripping water while you are spraying the window. If there are leaks, they will be detected earlier than they would be if you waited to see if the wall in the bedroom got wet.

If leaks are detected, you will need to locate the specific area where water is coming in. Direct the hose spray to one area at a time and cover the other areas with plastic sheets. Once found, correcting the problem may require that you remove and replace the window.

Damp Sash

I have wood-sash, double-glazed windows that have been stained and varnished. In the cooler parts of the house, condensation collects on the panes and drips onto the horizontal parts of the sash. Now the finish is starting to crack. What is the most water-resistant finish that can be applied to the sash?

For the best protection, the wooden sash should be coated with a polyurethane finish. Prior to applying such a finish, you should strip the sashes, sand them thoroughly, and then wipe them clean with a tack cloth.

When refinishing, apply the polyurethane so it covers about $1/32$ inch of the glass pane. This seals the joint between the glass and the wood.

Now that you've treated the symptom, it's time to correct the cause—excessive moisture buildup. First, determine where the problem originates.

There are a number of causes for excessive condensation, such as a crawlspace with no vapor barrier on the dirt floor, water seepage into the basement, or inadequate ventilation. An improperly operating humidifier or leaking air valves on steam radiators can also introduce excessive moisture into the house.

If you can't eliminate the cause of the high humidity, you can at least control it by ventilating rooms and using a dehumidifier.

Hard-to-open Windows

I have double-pane, double-sash windows in my house. They have been hard to open and close since the house was built. The builder said to apply soap to the window channels, but this does not make them slide any easier. I suspect the windows were improperly

installed. Perhaps they are not set square in their openings or are twisted slightly. I would appreciate any suggestions that can help me free these stuck windows.

There are several causes for stuck windows in relatively new houses. The most common is that an excess of fiberglass insulation has been packed between the window jamb and the window jack. This can also occur if too much foam insulation was squirted between the jamb and jack. Another likely cause is that the windows were racked slightly out of square when they were installed. Similarly, they may have been racked during shipment from the lumberyard and then not

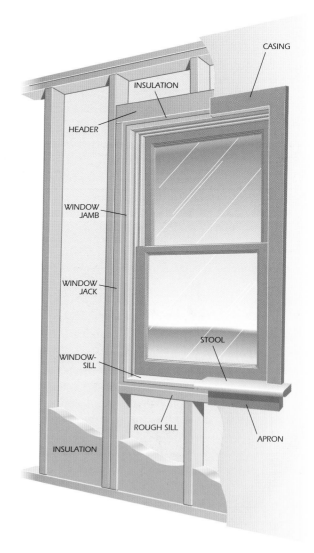

corrected when they were installed.

Occasionally, differential foundation settling shifts the framing around the window. This settling is easy to recognize because it causes cracks in the foundation wall and in drywall around door and window openings.

Finally, when excessively wet or poor-quality framing lumber dries, it forces the window jamb out of square.

To find the problem, you'll have to remove the window trim to expose the jamb. If you discover that the space between the window jamb and the jack has been overfilled with insulation, dig out the excess insulation with a long and sturdy screwdriver—a major undertaking if it has to be done to every window.

An inward-bowed window jamb can sometimes be pulled back slightly by driving a screw through the jamb and into the window jack.

Sash Problem

My house has new double-hung windows. The windows have spring-loaded sash cords. Three of the top sashes jam, and can't be released by pulling the cord back and forth. Also, when the sashes are pushed to the top and locked, they don't close tightly. What can be done about this?

There's a good chance that the window openings are not square. This condition can occur as the house settles, or be the result of improper installation. Also, when windows are installed, the voids between the jamb and framing should be filled with loose insulation. It's possible that an excessive amount of insulation was stuffed into the voids. This could push the jambs

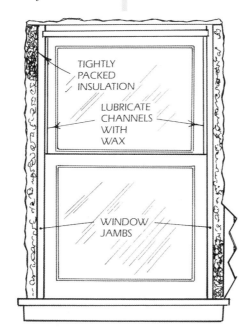

TIGHTLY PACKED INSULATION

LUBRICATE CHANNELS WITH WAX

WINDOW JAMBS

inward and reduce the clearance between the jambs and sashes.

First, try cleaning and lubricating the sash channels. You can lubricate them with furniture polish, paste wax, or candle wax rubbed in with a rag. Do not use oil. You can use silicone, but it must be a dry silicone spray—not a lubricant like WD-40.

If lubricating the channels doesn't correct the problem, then you'll have to remove the casing and resquare the window. This can usually be done from the inside.

Window Film

The ultraviolet film applied to the outside of our south-facing windows is cracking. We would like to remove it, but find that it's baked on. It's almost impossible to scrape off, and we worry about scratching the window. Can you recommend a way to remove the film?

I spoke with an authorized distributor of 3M Scotchtint Window Films about your problem. Based on his experience, he recommends that you mist the film with a spray bottle containing water and a few drops of a liquid dishwashing detergent. Apply two coats over the film when the sun is not shining on it. (Otherwise, the solution will evaporate rather than work its way through the film and into the adhesive below it.) Wait 10 to 15 minutes and then scrape the film using a 2-inch-wide straight-blade paint scraper. As long as you keep the blade flat, he says, you won't scratch the glass. It's a difficult job, but it can be done.

Solar Screening

We are interested in getting some information about solar screens for our windows. We've been told the screens block out 70 percent of the sun. How good are they, really? Summer temperature indexes of 110°F, measuring the combined effect of heat and humidity, are not unusual where we live. Also, should we screen all our windows?

Solar screens are similar to regular pull-down window shades, but are far more effective at preventing heat gain and sunlight damage. The screens are made of Mylar and are available in a range of colors, some of which are more effective than others at reducing heat transmission, fading and glare. The shades are quite effective and are custom-made.

Whether or not you install the shades on all windows depends on your taste and budget. If you are most interested in limiting heat transmission, then the windows facing south, west, and southwest should be shaded because these exposures receive the most amount of direct sunlight. Consider shading windows on northern and eastern exposures to reduce sunlight damage to furniture fabrics, or to your rugs or carpeting.

Storm-window Condensation

My house had wood double-hung windows and wood storm windows. Recently, we decided to do some remodeling. We kept the wood double-hung windows but changed the wood storm windows. I don't ever remember seeing condensation on my wood storm windows, but I see condensation on the inside (near the bottom) of the majority of my new aluminum storm windows. What can I do to prevent this from happening?

The condensation problem is not caused by the aluminum storm windows, but is rather the result of loose joints around the primary windows. In all probability, the wood storm windows that you replaced also had loose joints. Consequently, the warm, moist air that leaked through the primary window joints also leaked through the storm-window joints to the outside.

It will be very difficult to completely prevent condensation, but you can reduce it by using a removable caulk or weather seal around the inside of the windows and by installing weatherstripping.

Any caulk or weather seal will render the window temporarily inoperable, and you may want to consider whether this is more desirable than having foggy storm windows. Also, keep the weep holes at the bottom of the storm-window frame clean to allow any moisture that enters the area to escape.

DOORS

Condensation on Sliding Doors

Condensation forms during the winter on the inside of my sliding glass doors. How can I overcome this problem?

Apparently, the humidity in your house is quite high and the moisture in the air is condensing on the cold surface of the sliding door. Assuming that you don't want to replace the door with a more thermally efficient one, you should do the following:

First, check the humidity in your house with a hygrometer. An indoor relative humidity of around 35% is recommended for optimal comfort during the heating season.

Next, try to locate the cause of the moisture buildup. You may have a wet basement or a dirt-floor crawlspace that is moist. Also, be aware that moisture vapor is a byproduct of combustion, so have the heating system chimney checked to be sure that it's not plugged. If condensation persists, you could cover the door with plastic sheets on a removable wood frame.

Peeling door

For the last four years, we've had a problem with our energy-efficient foam-core metal door. After some cold weather, its paint peeled down to bare metal. We were advised to sand the door, remove any rust, apply several coats of metal primer and then paint, all of which we have done. It looks like we'll have to repeat the process, as it's rusting again. Any advice?

It's possible that either the door wasn't prepared properly prior to priming or the wrong primer was used. It's important the door be sanded down to bare metal. If you sandblast the door to strip off the paint and rust, then you must clean it afterward with a rag dipped in mineral spirits. The oil from the compressor could pass through the sandblaster's nozzle, leaving a thin film of oil on the door.

After preparing the door, apply a primer coat of zinc chromate. Don't wait a day to apply primer, because a thin layer of rust can develop. Let the primer dry for 24 hours, then apply a coat of exterior-grade, marine-quality alkyd enamel.

Split Door Panel

I am writing to find out if you have anything in your bag of tricks for fixing split exterior door panels. In our block of homes, there are about 8 out of 10 that are in this condition.

If the homes on your street are relatively new, the doors may be covered by the manufacturer's warranty.

The type of door that you are referring to is a stile-and-rail design consisting of vertical and horizontal sections separated by panels. Cracked panels are typical problems for solid wood doors, as opposed to veneer-covered particle-board doors.

The culprits that cause the door panels to crack are sunlight and moisture, both of which can be reduced with a long roof overhang.

Wood shrinks as it dries and it swells as it absorbs moisture. Changes in moisture content cause dimensional changes, splitting, cracking, and glue-line failures, especially if the wooden panel is wide, thin, and restrained.

It's important to reduce the amount of moisture absorbed by the exterior door to minimize the potential for panel cracking. This can be done by sealing and finishing the door, with either an exterior enamel or spar varnish. Apply the finish not only on the front, back and side edges but also on the top and bottom. The finish must be maintained, and the door periodically refinished.

DOORS

In order to further minimize moisture absorption, one door manufacturer mentioned that you can seal the perimeter joints of the door panels with a small bead of silicone caulk. There is not much you can do other than use a wood filler to cover the crack and then refinish the door.

Hollow-core Door

I have two hollow-core lauan doors in my house which lead from heated to unheated areas. I would like to insulate these doors by injecting some insulation through a series of holes bored in the hinge side of the door and plugging the holes with dowels.

I can't determine the best insulation for this use. Foam insulation in spray cans is one possibility; however, one door is 36 inches wide, and I wonder if the foam would reach the lower areas. Vermiculite is another option. Your ideas would be greatly appreciated.

You can't insulate hollow-core doors the way you suggest. Such doors don't have a large void framed between the face panels. For strength and stability, the panels (which are only about 1/10 inch thick) are bonded to one of three types of wood or wood-derivative cores: mesh (also called cellular), ladder, and implanted blanks.

Mesh cores consist of interlocked strips, which form a grid throughout the core. Sometimes honeycomb cores are used rather than a grid.

Ladder cores are made of vertical or horizontal strips which are roughly parallel.

Implanted blank cores consist of various shapes: spirals, tubes, or honeycombs, that may or may not be joined together.

Therefore, injecting foam or pouring insulation will not fill the core. Instead, try covering one or both faces with rigid insulation. For a pleasing appearance, the insulation can be covered with wall paneling.

Smooth Sliders

I live in a condominium that is 17 years old. The glass and screen doors have become a major problem, and owners are getting conflicting information on how to keep them opening and closing smoothly. The doors lift up and pull out, but most of us are not strong enough to do this. What is the best way to maintain these?

Once the doors are no longer sliding smoothly, get someone to lift them out of their track. Then clean around the bottom wheels. Over the years, dog hair, dust, and dirt accumulate around the wheel bearings, constricting them.

If the doors slide smoothly, the best way to maintain them is to keep the

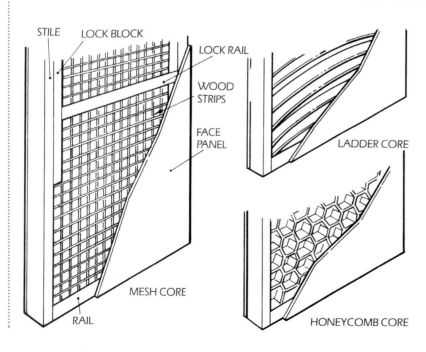

STILE LOCK BLOCK

LOCK RAIL

WOOD STRIPS

FACE PANEL

LADDER CORE

MESH CORE

RAIL

HONEYCOMB CORE

channel track that they slide in clean by periodically brushing it out or vacuuming it with a narrow nozzle. Wipe out any remaining grit with a damp sponge, dry the channel, then lubricate it with powdered graphite or a silicone spray.

Corroding Aluminum

The aluminum sills on my sliding glass doors are corroding. In some spots, the corrosion has eaten through the sills. The sills are set directly on a concrete slab. What can I do to stop the corrosion?

The aluminum sills should never have been placed in contact with concrete. An unprotected aluminum surface in contact with concrete and water a very corrosive environment. One way to isolate the aluminum from the concrete is to coat its bottom with a bituminous mastic, such as roofing cement.

The sill could also be placed on a strip of tar or asphalt felt. Any noncorrosive material can be used to isolate the aluminum sill from the concrete. Thin ($^3/_4$ inch or less) redwood or cedar boards can be used because they are rot resistant, and their thickness will not create a tripping hazard. Pressure-treated wood is not recommended for this application, because the salts used in the pressure treatment are corrosive to aluminum.

Cleaning Aluminum

The finish on the frames surrounding our 30-year-old storm doors has become dull and pitted. Is there a way to clean and polish them?

In all probability, the frames on your storm doors are fabricated with aluminum that has an anodized finish. An anodized finish is a glasslike oxide coating produced by an electrochemical treatment. The finish can be anywhere from .05 mil to 1.5 mils thick (1 mil is .001 of an inch).

Anodized surfaces are durable and weather- resistant, but they will weather over the years. Once the finish has become dull and darkened, it is difficult to bring back the shine. To some extent, the abrasive cleaners available through your local supermarket will clean and restore weathered anodized aluminum, but they cannot level pitted areas. And, although an anodized surface is tough, it can be damaged and even removed by vigorously applying an abrasive cleaner. When working with them it's important to follow the directions on the container and to test the product on a less-visible area first. Furthermore, it would be best to try a gentle cleaner, such as Bon Ami, before proceeding to one that is more aggressive. Also, try using a soft cloth in the initial cleaning process, as opposed to using a plastic abrasive pad. Don't use a steel wool pad. These pads may embed small steel particles in the aluminum. Although the finish will look fine at first, over time the steel particles will react with the aluminum and cause additional corrosion and discoloration.

After the door frame is clean and dry, you can provide protection by applying a coat of any kind of wax that is free of abrasives and cleaners.

5. Lot and Landscaping

DECKS • FENCES • DRAINAGE • RETAINING WALLS
FREESTANDING WALLS • BIRDS

DECKS

Deck Treatment

I am planning on power-washing my pressure-treated wood deck to remove accumulated dirt and mildew. Is it necessary to coat the deck after power-washing it?

Yes, the deck should be coated. Unprotected wood exposed to the weather goes through a swell-shrink cycle with each rain. This causes internal stress at the interface between the wet and dry wood cells, which eventually results in cracking, checking, and splintering. In addition, the lignin that holds the wood cells together degrades when it is exposed to the ultraviolet rays in sunlight. All of these processes are minimized when you coat the deck with lightly pigmented water repellent.

In order to control the growth of surface mildew, which results in gray and black stains, a mildewcide is added to the water repellent, and the mixture is then referred to as a water-repellent preservative.

I built a wood deck with treated southern pine. Can it be stained now, or must I wait a specific length of time? Which stains would be most compatible with it?

I'll assume the lumber you used was pressure treated with Chromated Copper Arsenate (CCA). This is the formulation used in Wolmanized wood (a trademark of the Koppers Company and not a generic term for pressure-treated lumber). These chemicals cause the familiar green tint in the lumber.

If the lumber's treatment stamp has "dry" included

in it, the wood was redried after treatment to a moisture content of 19 percent. Then, as long as weather conditions permit, you can paint or stain immediately.

More often, the lumber has not been redried. In this case, you should let it dry for 12 weeks before applying a semitransparent stain with an alkyd base that has a water repellent in it.

Treating Southern Pine

I want to stain my new southern yellow pine deck. Should I first apply a water-repellent preservative?

You can apply a water-repellent preservative (WRP) or a semitransparent stain, but you should not apply both because an application of WRP will inhibit the absorption of the stain. Besides, many semitransparent deck stains are water-repellent preservatives with a small amount of pigment added to give them color. These products will be absorbed by the deck and will inhibit mildew growth and prevent water from soaking into the deck.

It's important not to apply a stain that is more heavily tinted than a semitransparent type. However, you can apply heavily pigmented stains on vertical surfaces, such as handrail balusters. People apply solid stains to their decks, but are usually disappointed when the finish performs poorly. Research by wood technologists and finish chemists indicates that heavily pigmented stains don't perform well on horizontal wood surfaces that are exposed to the weather.

Grease Stain on Wood Deck

Our clear-oil-finished natural redwood deck was stained by greasy raccoon footprints when the animal disassembled our gas grill one evening. Is there any product or method that will lift the grease out?

First, try a good washing with a mild detergent and water using a stiff-bristle brush. The stain is probably a mixture of carbon and grease To prevent a blotchy appearance, wash the entire deck while placing extra effort on the stained area. If that doesn't do it, try washing the area with a solvent such as mineral spirits or paint thinner. Although the solvent will cause the wood to darken when applied, it shouldn't be noticeable after it evaporates. However, as a precaution, it's a good idea to try it on a small, nonobvious area beforehand.

If, after washing the area, a slight stain remains, ignore it and let it weather. Often, after a year's exposure to the sun, rain, and snow, this type of stain on a horizontal surface bleaches or washes out naturally.

Painting Pressure-treated Wood

I'm thinking of using pine lumber on the construction of a deck. Is pine available in pressure-treated form? Can it be painted?

Pressure-treated lumber has basically the same physical characteristics as untreated lumber, in that it can be painted, stained, or left unfinished.

Pressure treating will not protect wood from the ravages of weather exposure such as cracking, checking, splintering, and raised grain. To protect against weather-related damage, you should periodically apply a lightly pigmented water repellent or paint the surface. If you intend to paint treated wood, then check to determine that it's dry. Otherwise, the finish will not adhere.

Various wood species can be pressure treated. Southern pine is commonly used because its cell structure is well suited to accept preservatives. Some Western species of wood require the manufacturer to puncture the board's lateral surfaces to allow the preservative to penetrate.

When buying pressure-treated wood, it's important to specify the purpose for which it is intended. Wood treated for above ground use will have a concentration of .25 pound of preservative per cubic foot of wood, but wood intended for ground contact will have a minimum of .40 pound per cubic foot. The information should be on a tag stapled to the lumber.

Composite Lumber

Where can I find the new lumber made from recycled milk jugs? Our local bus system uses this product for bus-stop benches. I would like to buy various sizes of this lumber for building sheds and other outdoor structures.

In the last few years-wood-plastic composite lumber, which is available in many lumberyards nationwide, has become very popular. Its popularity stems from the fact that it does not require waterproofing and resists damage from weather and insects. The composites, however, are not designed to perform as a one-for-one structural replacement for wood in all applications. They are intended for decks, handrails, fences, and other outdoor applications rather than for structural, load-bearing elements such as joists, beams, and posts.

When manufacturing the composite decking material, several companies mix equal amounts of waste wood fiber and reclaimed plastic. During the mixing process, the plastic coats the individual wood fibers. The resulting material has superior weather-resistant properties.

FENCES

Painting a Chain-Link Fence

I want to paint my galvanized chain-link fence because it's starting to show signs of rust. What paint should I use, and how should I apply it?

Before painting the fence you must prepare its surface by wire brushing the rusted area. If the area is large, use a wire brush wheel chucked into a powerful electric drill or a right-angle grinder equipped with a steel- wire cup brush. Wash off the loose dust before proceeding to paint the fence, which can be done using an oil-based primer and topcoat.

The best tool for applying the paint is a long nap roller (its package probably will say that the roller is best suited for rough surfaces). The longer the nap the better, because the roller's fibers will reach through and around the fence material, almost painting both sides at once. You still need to paint the fence from both sides to do a proper job, however. Use a brush on the fence rails and posts.

Cold Galvanizing

I was surprised to see, in a past Homeowner's Clinic, that you advocated an oil-based paint for use on a chain-link fence to protect it from rust. Have you not heard of cold-galvanizing paint?

I am familiar with cold galvanizing. However, in that particular case the oil-based paint was appropriate. For readers who are unfamiliar with them, cold-galvanizing paints are manufactured by a number of paint companies, and they are very effective in providing galvanic corrosion protection. The paints consist of pure zinc particles in a binder, and in the dried paint film the zinc content is about 93 to 95 percent by weight.

In order to be effective, cold-galvanizing paint must be applied to bare metal. Any material left on the surface, specifically old paint or rust, will negate the galvanic protection because it will prevent the zinc from contacting the metal surface. The product is best applied when the surface is slightly roughened. This can often be done by wire brushing or using a right-angle grinder. However, if the surface has a heavy rust layer, it must be cleaned by commercial-grade sandblasting.

DRAINAGE

Controlling Groundwater

There is a 32-year-old workshop at the edge of our property, at a point lower than our house. The shop's ground floor is flooded every time it rains. The building is at ground level and sits on a concrete slab. It has gutters and extended downspouts to direct water away from the building. How can this problem be solved?

The problem you are having is typical of what occurs when a ground level structure is located at the foot of a slope. The slope causes surface and subsurface water to flow toward the structure, and unless the water is diverted it seeps into the building through open joints and cracks in the foundation or floor slab.

There are several things you can do to prevent or reduce water intrusion. Make sure that gutters are clean, so they don't overflow during a rain. Also, the ground should slope away on all sides of the building for a distance of 6 to 10 feet. This prevents water from accumulating around the foundation, and it forms a depression in the ground (also called a swale) that collects surface water. The swale extends around the sides of the structure so that water is directed away from the building.

SWALE AND PIPE DRAINAGE

2" TOPSOIL MINIMUM

GEOTEXTILE FABRIC

PERFORATED PIPE DISCHARGES INTO SOLID PIPE

GRAVEL OR CRUSHED STONE

Water Seeping Under Slab Floor

My home is built on a concrete slab. About five years ago, we began to get water seeping under the slab and into the floor heating ducts after a long, hard rain or a sudden deluge. Water enters the ducts on the side where the ground slopes upward, then fills all the other ducts. Can you suggest any way to stop this water from entering the house?

The best way to control the problem is to intercept the water before it gets under the slab and direct it away from the house. But before you go to the expense of putting in a subsurface drainage system, check other possible solutions that are less expensive.

The ground immediately adjacent to the house should be graded so it slopes away, as shown in the illustrationon the next page. This will prevent rain water from accumulating around the foundation. It also forms a swale, or depression, to carry away surface water. If you do not have gutters to channel roof rain runoff away from the house, they should be installed.

Subsurface water can be controlled with drains at the building's footing. These usually are installed when the structure is built. However, older buildings often have footing drains that are plugged or broken.

If a swale doesn't solve the problem, consider installing a curtain drain, as shown in the diagram. Often referred to as a French drain, this is a trench containing a perforated pipe surrounded by gravel. The pipe extends about 10 feet beyond the sides of the building. Note that a geotextile fabric covers the gravel to prevent soil infiltration from above. The fabric is covered with a minimum of 2 inches of topsoil; preferably 3 inches.

A solid drainpipe is connected to the end of the perforated pipe. The solid pipe is not surrounded by gravel, and it runs to a point where it can discharge into a storm drain or onto the lawn away from the building.

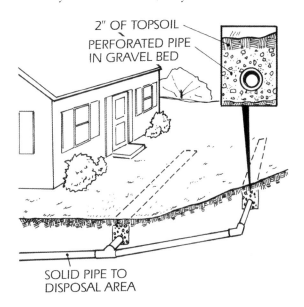

2" OF TOPSOIL

PERFORATED PIPE IN GRAVEL BED

SOLID PIPE TO DISPOSAL AREA

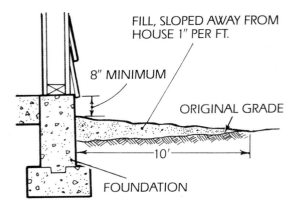

FILL, SLOPED AWAY FROM HOUSE 1" PER FT.

8" MINIMUM

ORIGINAL GRADE

10'

FOUNDATION

Drainpipe Backup

We have a drain line around the foundation of our house that runs to the edge of a stream. When it rains and the stream gets full, it covers the end of the drainpipe. Could this cause the water to back up in the drainpipe, or will it continue to drain out?

If you have gutters, make sure the downspout bases extend far enough to prevent runoff accumulation.

If surface water is not the cause of your problem, you will have to intercept the subsurface water movement. This can be done by installing a curtain drain parallel to the house on the slope that runs toward the house.

Dig a trench to a depth below the house foundation and fill it with a few inches of gravel. Lay a length of perforated pipe in the trench with the holes facing down. Fill the trench to within 2 inches of the surface with gravel and cover the gravel with topsoil.

Both ends of the perforated pipe should extend beyond the house. One end is capped, the other is connected to a nonperforated pipe, pitched to carry water to a disposal field.

It all depends on the elevation of the drain line relative to the stream when the stream is running full. A difference in elevation will create a pressure differential. As long as there is a greater pressure in the drainpipe, the water will flow out into the stream.

When the end of the drain line is submerged in the stream, water flowing down the drainpipe will flow into the stream. However, a residual amount of water equal to the level of the stream will remain in the drainpipe.

If the level of the stream, when the stream is running full, is above the level of the drain line that runs around the foundation, then water will back up into the drain line to the foundation and cause a water seepage problem.

DRAINPIPE BACKUP

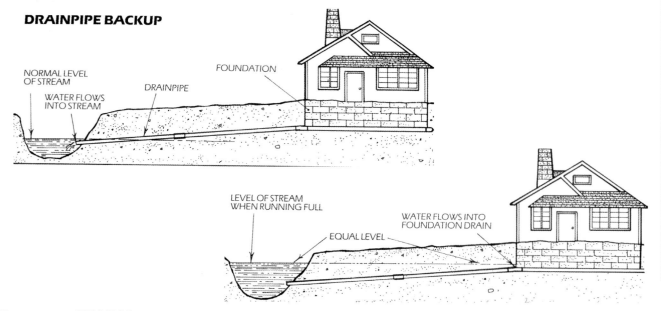

NORMAL LEVEL OF STREAM

WATER FLOWS INTO STREAM

DRAINPIPE

FOUNDATION

LEVEL OF STREAM WHEN RUNNING FULL

EQUAL LEVEL

WATER FLOWS INTO FOUNDATION DRAIN

RETAINING WALLS, FREESTANDING WALLS

RAILROAD TIE RETAINING WALL

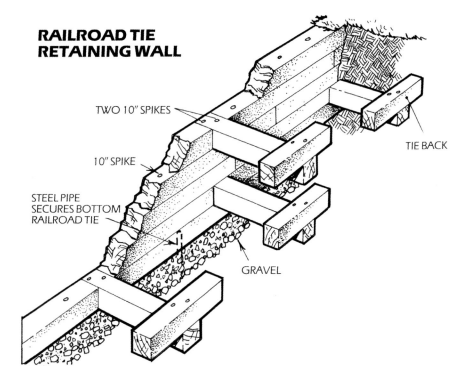

TWO 10" SPIKES

10" SPIKE

STEEL PIPE SECURES BOTTOM RAILROAD TIE

GRAVEL

TIE BACK

RETAINING WALLS, FREESTANDING WALLS

Rotting Railroad Ties

I built a retaining wall of used railroad ties. In a few areas, the rot is creating holes. Is there any way I can remedy this, besides rebuilding the entire wall?

There really isn't much you can do about repairing the rotting sections in your retaining wall. There are epoxy fillers that are used to rehabilitate rotting trim in houses, but they're not intended for retaining walls. If you have only a few sections that are rotted, and if the structural bracing for the wall (tiebacks) has not deteriorated, then rebuilding the entire wall should not be necessary. You can remove and replace the extensively rotted ties.

Rotten used railroad ties are not unusual. This is because the species of wood that is used for railroad ties is difficult to impregnate with creosote. Consequently, there is quite a lot of untreated wood and moisture in the interior of the ties. When the ties dry, they check and split, exposing the interior to decay.

Furthermore, creosote consists of many components, some of which are water soluble. Over time, the creosote retained in the tie is depleted as these components leach out. Eventually, decay occurs.

Stone Garden Wall

I'd like to build a freestanding stone garden wall in my rear yard. Can you give me some tips?

There are two types of stone wall construction: a dry wall (built without mortar) and a wet wall (built with mortar).

RETAINING WALLS, FREESTANDING WALLS

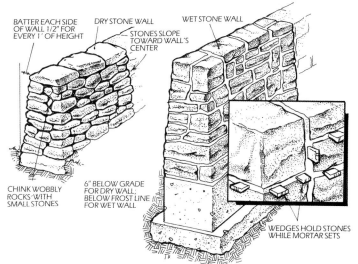

BATTER EACH SIDE OF WALL 1/2" FOR EVERY 1' OF HEIGHT

DRY STONE WALL

STONES SLOPE TOWARD WALL'S CENTER

WET STONE WALL

CHINK WOBBLY ROCKS WITH SMALL STONES

6" BELOW GRADE FOR DRY WALL; BELOW FROST LINE FOR WET WALL

WEDGES HOLD STONES WHILE MORTAR SETS

With a dry wall, the stones can be quickly restacked if the wall is damaged. This wall needs no footing because it floats with frost heave. The base of the wall should be about 6 inches below grade. Use the largest stones for the base. This avoids the need to lift and place them.

The mortar in a wet wall keeps the stones in place and makes the wall act like a monolithic structure. Such a wall needs a footing extending below the frost line to protect it from heaving caused by freeze/thaw cycles.

Freestanding walls are usually no more than 4 feet high, and should be inclined from the vertical (battered) at a rate of $1/2$ inch per foot of height.

Drainage for Concrete Block Wall

I have a painted concrete-block retaining wall on the property line with my neighbor. When it rains, excess moisture forms a white powder on the wall, and its paint flakes off. How can I treat my side of the wall to stop the moisture seepage and prevent the paint from flaking? We were advised to tar the wall's back. Drains were installed at the wall's base, but the clay in the soil seems to prevent the water from seeping down to the drains.

The white powder on the wall is efflorescence, a mineral deposit that forms when water that's carrying minerals evaporates. To prevent this, and stop the paint from peeling, you must stop the water from seeping through the wall by installing better drainage. Treating your side of the wall won't solve the problem.

You are correct in saying that the clay in the soil prevents the water from reaching the drain. Soil with a high clay content does not drain well. This soil should be excavated from behind the wall and replaced with gravel. A perforated drainpipe should be installed as shown. Water will percolate through to the pipe. It is important that the pipe have a free-flowing outlet to allow the accumulated water to drain.

Do not simply tar the wall's back and then backfill with the same soil—the tar will seal the wall and prevent the water in the soil from escaping to the outside. If the water builds up, it can exert enough pressure on the wall to crack it or heave it.

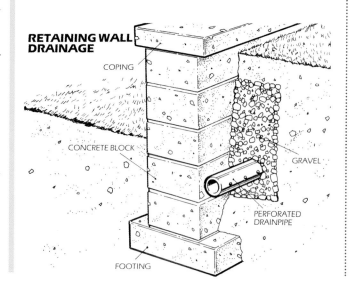

RETAINING WALL DRAINAGE

COPING

CONCRETE BLOCK

GRAVEL

PERFORATED DRAINPIPE

FOOTING

RETAINING WALLS, FREESTANDING WALLS

Spalling Brick

I have a brick wall about 3 feet high on either side of the driveway. Behind the wall is a grassy area. Quite a few of the bricks have spalled—the face has broken away, leaving cavities all over. Probably each broken brick will have to be chiseled out and a new brick inserted with mortar. What can I do?

Apparently water is being absorbed by the bricks. During winter this water freezes and thaws—a cyclical frost action that causes the bricks to spall.

Before replacing the deteriorated bricks, you should eliminate the source of the water. Seal and repoint all cracked and open joints in the coping on top of the wall, and also between the exposed bricks.

Perhaps the grassy area does not drain properly. If that soil periodically stays wet, it could also cause the spalling. In this case the area should be excavated and filled with gravel to a level about 6 inches below finished grade. Cover the gravel with polyethylene sheets and then with topsoil. Install weep holes in the bottom of the wall if there aren't any.

BIRDS

Bird Control

Can you recommend a method for preventing birds from landing and roosting on the ledge over the front entry door to my house? Each spring the birds build a nest on the ledge and bird droppings end up on the entry steps below. I don't want to harm the birds; I just don't want them to come back. What can I do before next spring to avoid a problem?

Although birds are a delight to watch, they can also be a nuisance and a health hazard. Their droppings are highly acidic and can discolor and damage masonry and metal. The droppings are also associated with a number of transmittable diseases—some of which can be fatal. Keeping the birds away from your doorstep is a good idea.

What you need is a bird prevention device that is approved by the Humane Society of the United States. One device, which is available at pest-control companies, consists of plastic spikes mounted at different angles on a 2-foot plastic base so that each section provides a 7-inch width of coverage. This is effective because the spikes are close enough to one another to deter the birds from landing. The sections can be easily installed, and are available in seven standard colors

SPIKE 2000

BIRDS

Birds versus Cherry Trees

I have two cherry trees in my front yard that I have to net each season so that the birds won't eat the fruit. Is there a way to protect the trees without nets, perhaps by using a high-tech device? I've wondered how big-time growers protect their trees.

There's no high-tech solution to the problem. I spoke to the local horticultural agent for the Cooperative Extension Service, who said that many commercial growers pick the fruit before it's ripe. Birds don't find the unripe fruit very appealing, so they leave it alone. The cherries ripen either during shipment or at the stores themselves.

The best approach is to cover the trees with netting. This discourages birds—and it won't harm the tree or the fruit. Some people use plastic owls and sound devices to scare off birds. This isn't as effective as nets, however, and some communities have passed noise ordinances that prohibit such devices. We even checked with the National Arbor Day Foundation to see if any chemical repellents could be sprayed on the trees. The NADF says there are none.

6. Garage–Shed

OVERHEAD DOORS • FLOOR SLAB/SEEPAGE
• MISCELLANEOUS • SHED

OVERHEAD DOORS

Electric Garage-door Opener

I have a 20-year old, electrically controlled garage-door opener. I can open the door from the car with a remote control or by using the key switch. Can I install a system that uses a numerical keypad instead of the switch? It seems like a better system than relying on a key, which can be forgotten inside the house or left elsewhere.

You can replace the key switch with a keypad. However, in your case, the keypad has to be wired into the opener. Newer systems do not require the keypad to be wired to the opener. They send a signal from the keypad to a receiver on the door's drive mechanism. Your opener will not work with this system.

A keypad is more convenient than a key switch and, in my opinion, is more secure. I once saw a demonstration in which a toy water pistol was used to activate a garage door opener. Water from the pistol was squirted into the keyhole and shorted the electrical connections inside, activating the opener.

In 1993, safety regulations were passed that require garage-door openers to have a photoelectric beam and sensor. The beam is projected across the door opening and if a person, animal, or object breaks the beam as the door is closing, the door stops and then opens. If your door-opener system does not have an automatic reversing mechanism, rather than just replacing the key switch, you should consider replacing the entire system.

Garage-door Control

My daughter has a 10-year-old garage-door opener that needs repair, but the repairman says the manufacturer told him replacement parts are no longer available. Can you help?

Even if the parts were available, the repairman might not want to install them because the door does not meet current safety standards. The garage door opener does not comply with recent guidelines established by Underwriters' Laboratory (UL Standard 235). This is important because, between 1982 and 1988, 48 children between the ages of 2 and 14 died from being trapped under garage doors that were operated by automatic openers.

Federal law mandated that as of January 1, 1993, all new residential garage doors must comply with UL 235. This requires that garage-door opener assemblies include a photoelectric sensor or other device that will reverse the door if it comes in contact with someone or something as it is closing. It also requires the apparatus to automatically disable the opener should the sensor malfunction.

There are no federal restrictions on repairing older residential garage-door openers, but several states require that these older devices be brought into compliance with UL 235 at the time repairs are made to them.

OVERHEAD DOORS

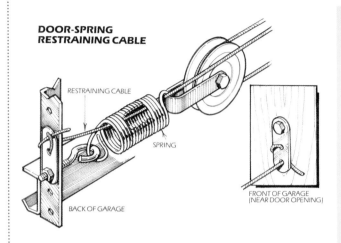

**DOOR-SPRING
RESTRAINING CABLE**

RESTRAINING CABLE

SPRING

BACK OF GARAGE

FRONT OF GARAGE
(NEAR DOOR OPENING)

Garage-door-spring-Restraining Cable

The spring on my roll-up overhead garage door recently broke. Luckily, I wasn't standing nearby and my car was out of the driveway. Otherwise, I could have been injured or my car damaged by the broken spring as it whipped by. The father of a friend of mine lost a fingertip because of a broken garage-door spring. Since a spring could snap at any time, is there any way to prevent an injury or serious damage?

When a garage door spring breaks, the loose end whips around with considerable energy. To prevent this, install a restraining cable through the center of each spring. This prevents the spring from whipping around. The cable is secured at each end with plates that are bolted down, and it will not interfere with the action of the spring or the door.

Cables can be purchased from a garage door service company. Restraining cables are inexpensive, a worthwhile safety feature, and a good do-it-yourself project.

Wobbling Garage Door

My single-car garage has a heavy wood garage door that wobbles when it is opened or closed using the door's automatic opener. Is there a simple remedy, or do I need a new door?

Several things can cause a garage door to wobble when opening or closing it. It's possible that the tracks could be out of alignment, or the rollers may have flat spots. Also, if the door has springs, they may have uneven tension.

A door is installed in sections, and if it's not assembled correctly, then the joints between the sections are not parallel. I doubt that you need a new door. However, you could use the services of a garage-door mechanic.

If the garage door has springs, have the mechanic install a restraining cable inside each spring after the wobble is corrected. This cable holds the spring in place if the spring breaks while it's under tension. This prevents the broken spring from flying through the air and causing damage or injury.

FLOOR SLAB/SEEPAGE

Peeled Garage Floor

My problem is the peeling paint on our concrete garage floor. The floor was dirty and lumpy. We cleaned it and the ground areas smooth, then applied deck paint. It looked lovely until the car left black tire marks on it. Then we cleaned it, and now it's peeling. Can we fix it?

Three factors cause garage floor paint to perform poorly. They are improper surface preparation, moisture intrusion, and hot-tire pickup. All three can act together to cause the paint to peel. First, it's difficult to adequately clean a garage floor because dirt and grease are ground into the concrete's pores over a period of many years. Also, the floor is usually cracked, and most floors lack a vapor barrier below the slab. Because of this subslab ground moisture works its way to the slab surface and lifts the poorly bonded paint from the slab surface.

The porosity of latex paint helps it resist peeling from moisture vapor, but the main problem with latex paint is hot-tire pickup. Car tires get quite hot, and when the car returns to the garage, the hot tires soften the paint on a garage floor and cause it to stick to the tires. Oil paint resists hot-tire pickup, but subslab moisture causes it to peel more readily.

Given the number of problems that can occur with paint on a garage floor, I don't recommend painting it. A better alternative is to clean the floor with a commercial concrete cleaner or high-pressure washing, or both, and apply a masonry stain. The stain allows moisture vapor to pass through it and it is less likely to peel or be lifted by car tires.

Concrete Topping

I've been informed that there is a process of applying a thin layer of topping material to a concrete garage floor. I would appreciate any assistance regarding this.

You can top a slab, but you must be careful not to get the topping material too thin. The Portland Cement Association recommends that topping over hardened concrete should not be less than 2 inches thick at any point.

To top a concrete slab, clean the surface with muriatic acid or concrete cleaner (available in paint and hardware stores). To ensure that the topping bonds well, roughen the surface of any slab that appears shiny and smooth. Use a wire brush and full-strength muriatic acid to do this.

Next, mix a slurry of cement, sand, and water and scrub this onto the surface with a throwaway scrub brush. This slurry acts as a primer to ensure that a good bond forms between the slab and the topping. Apply the concrete topping mix on top of the slurry before the slurry starts to dry.

Observe standard concrete-work procedures when topping the slab. Keep the topping moist as it cures by covering it with burlap or misting it. Also, don't work the top to a very smooth finish. A rough surface is less slippery when it's wet.

Floor Slab Leak

I have water leaking through the concrete-slab floor of my garage. What kind of grout or patching cement will stop the leak?

The leakage can be corrected by sealing the cracks and any open joints with a nonshrink hydraulic cement, which is generally available at hardware stores.

Before applying the cement, prepare the cracks or open joints by undercutting or square cutting. Do not use a V cut. Add enough water to the cement to get a putty consistency and then force it into the crack with a trowel or gloved hand. At floor-to-wall joints, form the cement into a cove for increased effectiveness.

If the seepage is heavy, it could be the result of a high water table beneath the slab. Sealing the floor could result in excessive hydrostatic pressure that might cause the slab to heave and crack. In this case, the best solution is to lower the level of the subsurface water by installing perimeter drains below the slab and running them to a sump pit where the water can be pumped away.

Water Seeps In

When we get a big downpour my garage fills up with water. The water seeps up through the concrete floor slab. Is there anything I can do about it?

The first thing you should do is inspect the area outside your garage. If it has gutters and downspouts (and it should), they should discharge away from the house. Also, the ground adjacent to the garage should slope away from it so that surface water will not flow toward the foundation and accumulate under the garage floor slab.

If neither the downspouts nor ground slope cause the problem, then you will have to install a sump pump in the floor slab to control the level of subsurface water.

If the drainage below the slab is good, then a single pump should be adequate. With good drainage, water will flow through the gravel bed below the slab from the farthest corner of the garage to the sump pit. However, if the drainage is poor, water will not flow readily beneath the slab. In this case, a series of connected perforated pipes that discharge into the sump pit should be installed below the garage floor. This is a major undertaking. The outline of each trench is sawed into the floor. Then the concrete between the saw kerfs is removed with a jackhammer. The process creates considerable noise and dust.

Patching Spalled Concrete

The concrete floor in my garage has areas that flake away as I sweep it, and the rough stones are protruding. I've tried using liquid sealers, including vinyl cement, but they don't hold. Can you help with this problem?

Spalled sections on a concrete garage slab are generally caused by deicing salts dripping from the car. There are a number of materials for patching. If epoxy or polymer patches have not held, chances are you didn't prepare the surface properly. The area to be patched should be roughened by sandblasting or chipping to about 12 inches beyond the spall.

The area should be clean and dry. Remove oil and grease, loose concrete and dust. Make sure there is no moisture rising through the concrete. Since epoxy products vary, follow the manufacturer's directions.

Depending on the size of the spalled area, it may be more economical to use a concrete mix. However, this will require more surface preparation since concrete cannot be "feathered" out like epoxy. The edges of the area to be patched must be undercut to hold the mix in place. The procedure outlined below for patching a floor is from the American Concrete Institute's booklet "Slabs on Grade."

1. Using a circular saw with masonry (silicon-carbide) blade, make a rectangle of outward-angled, $1/2$-inch-deep cuts around the spalled area.

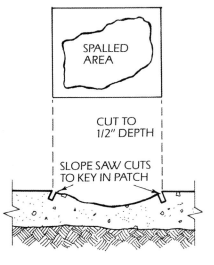

SPALLED AREA

CUT TO 1/2" DEPTH

SLOPE SAW CUTS TO KEY IN PATCH

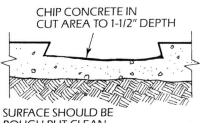

CHIP CONCRETE IN CUT AREA TO 1-1/2" DEPTH

SURFACE SHOULD BE ROUGH BUT CLEAN

2. Wearing goggles, chip concrete within the saw-cut area to about $1^1/_2$ inches deep using a cold chisel. Chipped surface should be rough but clean.

3. Dampen area with water, cover with wet burlap, and allow it to stand for several hours.

4. Mix concrete for patch in the same proportions as the slab. If you don't know the mix of the original concrete, use a ratio of 5.5 gallons of water per 94-pound bag of cement. Let the mix stand several minutes before placing it in the patch.

5. Remove excess water from the patch area, but leave surface damp.

6. Compact concrete into the patch, overfilling slightly.

7. After a few minutes, level it to match the surrounding surface, then finish to the required texture.

8. Keep the patch damp for three days.

MISCELLANEOUS

Gasoline Vapors Pose Explosion Hazard

I recently inspected two condominium townhouses in two municipalities of Westchester County, New York, with a condition that, in my opinion, is an explosion waiting to happen.

Both townhouses have a gas-fired heating system in the garage (one is a forced hotwater boiler, the other a horizontal furnace). Both appliances were about 4 inches above the floor.

A garage is a fire-hazard area because of the presence of gasoline, paints, and solvents. If spilled, paints and solvents (or leaking gasoline) can release vapors that could explode when ignited by a pilot light or a gas burner that is firing.

According to the National Fuel Gas Code (NFGC) and the New York State Building Code, gas-burning appliances located in a garage in a one-or two-family dwelling shall be installed so the burners and ignition device are not less than 18 inches above the floor. The reason that the heating units were allowed to be installed with the burners only 4 inches off the floor, and not 18 inches, is because the problem fell through a bureaucratic "crack."

Condominium townhouses are not classified as one- or two-family dwellings but as multifamily dwellings. The state building code does not specifically address the above condition for multifamily dwellings, so local code officials accepted the installations.

If this condition exists where you live, you should isolate the heating unit or build an 18-inch high masonry wall in front of it.

Save That Garage

I've bought an old house with a small detached garage. Rain runoff flows down my driveway and through the garage, and the sole plate has been repeatedly soaked. Both the sole plate and the bottoms of the wall stud have started to rot. The rest of the building is in good shape and I would like to save it for a workshop. I'd appreciate any help.

Before repairing the garage, you should eliminate the water penetration. Install a drain across the driveway in front of the garage to catch and deflect the runoff. Cut a small channel, about 8 inches wide across the driveway, fill it with gravel, and cover it with a grate. Provide a free-flowing outlet using 3- or 4-inch-diameter pipe to direct the water downhill and away front the garage.

If the driveway is not steep, you might simply divert the water with an asphalt lip across the driveway, 2 to 3 inches high. Diverted water should flow to a lower area in the lawn.

Before cutting away the rotted portions of wall, you must erect supporting braces. Working on one wall at a time, nail a 2 x 4 horizontally along the top inside edge of the wall with the wide face against the studs. Wedge a

2 x 4 under every other ceiling beam, between the horizontal 2 x 4 and the floor. This will relieve the pressure on the wall and allow the rotted framing to be cut away.

Remove the rotted sill and cut off the bottom of the studs 11⅛ inches above the floor. Set a row of 8-inch concrete blocks (that actually measure about 7⅝ inches high) in a ½-inch bed of mortar so that they align with the outside of the garage wall.

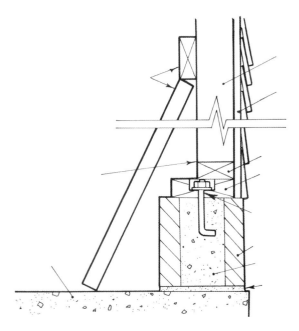

Mortar anchor bolts into the block cavities so they protrude 1½ inches above the top of the blocks. Use three anchor bolts per wall. Bolt a 2 x 6 to the blocks with holes counterbored for washers and anchor nuts. Then, nail a 2 x 4 along the length of the 2 x 6 to form a sole plate. Finally, toe-nail the studs to the 2 x 4 with galvanized 8d nails.

Cold Floor

The hardwood floor of my living room is unbearably cold in the winter. Below the living room is the garage, which has a ceiling with 6 inches of insulation between the joists, held in place with a nylon netting. How could I better insulate the garage ceiling to solve my problem of "cold feet?"

You should be more concerned with the potential fire hazard in the garage than with the cold floor. Nevertheless, minimizing the former will also help the latter. An attached interior garage that is used for parking a car is considered a fire hazard because of the gasoline in the car. Exposed floor joists should be covered with fire-rated gypsum drywall, such as Type X sheetrock.

If a fire should develop in the garage, the exposed wood framing, such as the overhead living room floor joists, would quickly burn, causing the living room floor to collapse. With Type X sheetrock covering the exposed wood, it would take longer for a fire to burn through the floor. This provides more time to escape.

The exposed floor joists are responsible for the cold floor. Although wood joists are not good heat conductors, they will conduct heat away from the floor. To reduce this—and the fire hazard—cover the joists with rigid insulation boards, followed by fire-rated drywall.

SHED

Concrete-slab Reinforcement

I am getting ready to pour a concrete slab that will be the foundation and floor for a work shed. I'm considering using rebars to strengthen the slab, but I don't know how far apart to space them. The shed measures 10 x 14 feet. Is there a rule of thumb that I can follow?

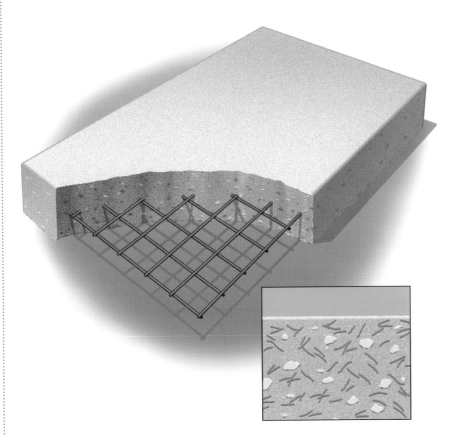

Although there is no universally accepted rule of thumb, some engineers consider a minimum practical spacing for rebar in a concrete slab to be approximately equal to the slab thickness, and a maximum spacing between bars to be three times the thickness.

In practice, rebar spacing is calculated based on code requirements, for which a number of variables are considered, such as the loads to be imposed on the slab, the soil's bearing capacity, and the likelihood of frost heaves. These variables give engineers a rough idea of how thick the slab should be, the type of concrete to specify, the diameter of the rebars to be installed and how far apart to space them. Then they perform the necessary engineering calculations to arrive at specific rebar spacing and slab design.

In your case, the slab will be supporting only a small load, and it will be installed entirely on grade. In these types of installations, the slab is usually reinforced with welded-wire mesh, also known as welded-wire fabric. A common type has 6-inch-square holes.

A relatively new method of reinforcing concrete is to add polypropylene fibers or other types of fiber to the mix. The fibers form a chemical-mechanical bond with the surrounding concrete, and this results in a three-dimensional reinforcing system.

7. Wood-destroying Insects

TERMITES • POWDER POST BEETLES • MOLES

TERMITES

Spotty Termite Treatment

A pest-control service spot treated my home for termites. But it didn't dig any trenches to treat subterranean termites, and it seemed like the service finished the entire job too quickly. I am questioning the quality of the treatment.

If your house has never been treated for subterranean termites, and an active termite condition is found, then the entire house should be chemically treated with termiticide. This is important because termite activity oftentimes is not limited to one area, and the insects may be in areas that are not visible, such as the cavities of a concrete-block foundation. Treatment generally includes trenching or injecting the chemicals into the ground around the foundation.

However, if the house has already been treated and an active condition is discovered, then most state environmental regulations allow only a spot treatment.

A Case of Mistaken Identity

I recently noticed sawdust on the floor below a wood beam in my basement. Is this an indication of a termite condition?

No, it's more likely an indication of carpenter ant activity. Carpenter ants, unlike termites, do not eat

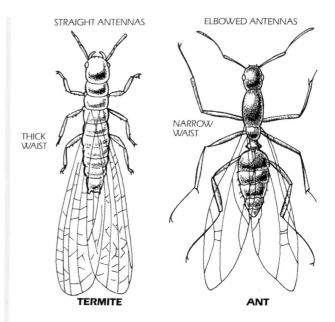

STRAIGHT ANTENNAS — ELBOWED ANTENNAS
THICK WAIST — NARROW WAIST
TERMITE — **ANT**

wood. They merely excavate it to build a nest. The small fragments of shredded wood they generate during the excavation are removed by the ants and deposited outside the excavation. Termites, however, completely devour wood they are attacking and leave no wood particles behind.

Both carpenter ants and termites have a rigid caste system within their colonies. Reproductive insects from these colonies periodically sprout wings and fly off to set up new colonies (swarming). The two insects look alike, but the most recognizable difference between them is that the ants have a pinched waistline.

A carpenter ant infestation can be eliminated only by destroying the nest, which can sometimes be located by watching the ant traffic. Once found, it can be destroyed with insecticide. If it can't be found, individual ants can be controlled with dust or spray insecticides. Remember to exercise caution when using insecticides, and follow the manufacturers' directions.

POWDER POST BEETLES

Infested Joists

The 65-year-old-house I recently purchased has some bad floor joists. New wood framing members were placed beside the old. The old floor joists are peppered with $\frac{1}{8}$-inch holes. They are thoroughly tunneled and full of wood dust. I tore down some of the worst floor joists with my hands and saw several insects that look like silverfish. What can I apply to kill the insects and prevent further damage? Someone recommended coating the framing with used motor oil.

Your description sounds like the work of powder post beetles rather than silverfish. Silverfish, though a common pest, do not eat or damage wood. Powder post beetles, however, do.

The beetles lay their eggs in cracks and crevices in the surface of unfinished wood. After the eggs hatch, the larvae feed and tunnel through the wood, reducing it to a powdery residue. Just prior to emerging from the wood, the newly formed adult beetles chew small round holes ($\frac{3}{32}$- to $\frac{1}{8}$-inch diameter) in the wood surface. Shortly after emerging, the beetles mate and lay eggs, occasionally depositing them in the opening of an old exit hole, and reinfesting the same piece of wood.

Those joists that have extensive damage have been infested by several generations of powder post beetles. It's very possible that the floor joists were infested before being used in constructing your house. Definitely do not use motor oil to coat the joists. The motor oil would create a fire hazard and produce unpleasant odors in your basement.

Controlling active infestation is best handled by a licensed pest-control operator using either liquid treatment or fumigation. Both methods present potential environmental hazards and should be administered by a professional.

MOLES

Moles in Lawn

Moles moved into my yard last summer and they are already working this year. Please tell me how to get rid of them.

There are two methods for ridding your yard of moles. One is to eliminate their food source, which normally consists of earthworms and beetles. To do this, you need to apply an insecticide. But in ridding your lawn of earthworms, which are beneficial to the lawn's health, you may do more harm than good.

The second method is to trap the moles. Some traps kill the moles, while others leave them essentially unharmed for release elsewhere. The best time to trap moles is early spring, as soon as you see the ridges that are their runways forming in the lawn. You need to determine which runways are active. To do this, stamp down a section of runway and check it over a period of several days. If the ridge is reformed, stamp it down again. If it is raised again, the runway is active, and a trap should be set at that location.

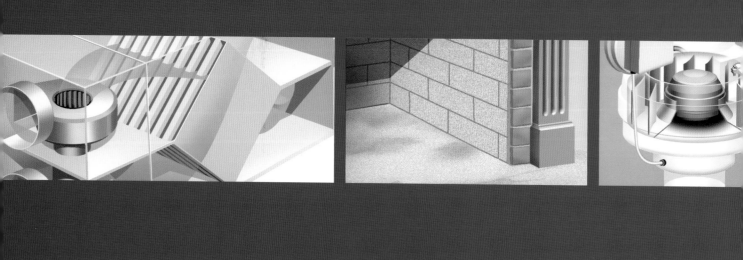

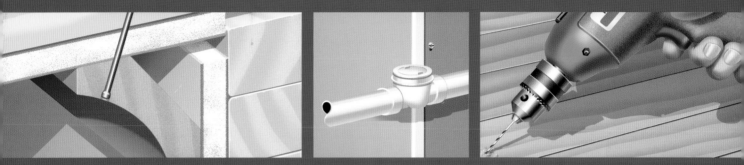

Interior

8. Attic

INSULATION ● VENTILATION/MOISTURE ● ATTIC FANS ● TRUSS UPLIFT ● MISCELLANEOUS

INSULATION

Drafty Folding Stairway

I have a problem with the folding stairway to my attic. My furnace thermostat and return duct for the furnace and air conditioner are located in the same hallway as the stairway. In the winter, a cold draft comes down through the cracks around the stairway and causes the furnace to come on even though the rest of the house is warm. In the summer, the reverse is true.

An easy and effective solution is to construct a rigid insulation cover for the stairway. In addition to reducing the draft, the cover will reduce the heat loss through that area.

The problem that you describe is quite common. I've inspected thousands of homes with folding stairways used for attic access and over 95 percent of them did not have insulation over the stairway opening.

You can use an easily available rigid foam-insulation board, such as Styrofoam, to construct your insulation cover. The top of the cover should be large enough to overlap the perimeter of the stairway opening by a couple of inches. The sides must be deep enough to accommodate the folding stairway in its closed position. Once the top and four sides are cut to size, they can be attached to one another with an adhesive.

The solvent in some adhesives will react with Styrofoam. Be sure that the product you buy is compatible with the material you're gluing.

A 1-inch-thick Styrofoam board has a thermal resistance (R-factor) of about 5. If you'd like greater thermal resistance, then either double up on the Styrofoam or insulate the cover with Fiberglas batts. Since the cover will weigh only a few pounds, it can easily be moved around when you use the stairway.

Attic Insulation

When we bought our 1935 farmhouse a friend told us that we could not insulate our attic ceiling because there is already insulation there with a vapor barrier. The insulation is between the second-floor ceiling and the attic's hardwood floor. Is it true that we cannot insulate and finish the attic? It's a full walk-up attic. It would be a shame to waste the space.

It's always possible to add insulation to an attic, even if there is already insulation with a vapor barrier in the floor. The only questions are where to place the new insulation and whether it should have a vapor barrier.

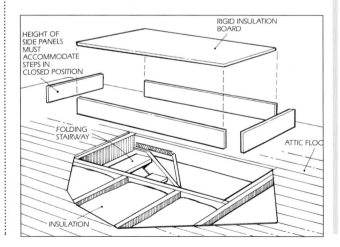

HEIGHT OF SIDE PANELS MUST ACCOMMODATE STEPS IN CLOSED POSITION

RIGID INSULATION BOARD

FOLDING STAIRWAY

ATTIC FLOOR

INSULATION

First, some basic information on insulation is in order. In an unheated and unfinished attic, insulation is installed between the ceiling joists, and its vapor barrier should face the warm room below. Assuming that the vapor barrier in your case was installed facing the room below, the insulation was correctly installed. When further insulation is installed on the floor of an unfinished attic, it should not have a vapor barrier. This second layer of, insulation may be installed directly on top of or at right angles to, the insulation below it. If insulation with a vapor barrier is installed, the vapor barrier may form a surface on which moisture vapor will condense. Wet insulation is not only ineffective, it also forms an environment conducive to wood rot and insect habitation.

When an attic is finished, there's no need to install more insulation in the floor unless you are trying to achieve sound-deadening with it.

A finished attic is treated like the portions of the house below. The insulation is placed with the vapor barrier facing the heated and finished space inside the attic. In this case (and most other cases) this requires that insulation with a vapor barrier be placed between the rafters and between studs on the end walls. It's extremely important, however, that the insulation installed between the rafters not press directly against the roof deck. An airspace must be left above the insulation. It's best to install soffit and ridge vents so that air can move from the soffit, behind the insulation and out the ridge. This will keep moisture vapor from collecting and condensing on the underside of the roof deck.

The position of the vapor barrier is a source of confusion to homeowners. The vapor barrier is placed on the warm side of the structure so that moisture vapor will not condense on it. A vapor barrier essentially stops the passage of moisture vapor. In doing so, the moisture vapor does not condense in wall or roof cavities. If the vapor barrier is placed on the cold side of the structure, then moisture vapor from living areas will pass nearly unrestricted through the insulation and condense on the back of the vapor barrier.

There are a few other things that you should consider when finishing an attic, especially one in an older house. Attics in old houses frequently have fragile and outdated electrical wiring and electrical devices installed in them. Have an electrician inspect the attic before construction begins. Finally, check with the town's building department to see if a building permit is required to finish the attic.

Foil Radiant Barrier

I would like to install a foil radiant barrier in my attic. There are batts of insulation between the ceiling joists with another 2 inches of blown-in insulation on top of that. I need the foil barrier more to keep the attic cool than for heat, and I would like to lay it on top of the insulation. Can I install it this way?

For readers not familiar with them, a foil-faced radiant barrier reflects radiant heat from the sun away from the attic, reducing the amount of air conditioning necessary to cool the rooms below.

Don't install the barrier on the existing floor insulation, because in most cases it acts as a vapor barrier and causes condensation problems. Instead, the barrier should be attached to the underside of the roof rafters. The joints should not be sealed, so that moisture accumulating in the cavity between the roof deck and the radiant foil barrier can escape.

If only one side of the foil barrier is shiny, install that side face down. This prevents dust from settling on the surface and lowering the barrier's reflective capability. According to tests, dust accumulation can reduce the barrier's performance by 50 percent.

To allow for air circulation behind the barrier, leave a gap at the bottom and top of the rafters. The industry trade group that oversees radiant barriers specifies a 3-inch gap at the eaves and a 3-inch gap on each side at the ridge (6-inch total gap at the ridge). Finally, install a radiant barrier according to the manufacturer's instructions.

Asbestos Information

I am currently remodeling my 42-year-old house. Much of the blown-in insulation is processed recycled newspaper. When I started removing some of the insulation, I found a material underneath it that looks like cotton balls, and I'm wondering if it's asbestos. Is it possible for a homeowner to identify asbestos based on its appearance?

You cannot identify asbestos based solely on appearance. To positively identify asbestos, a bulk sample of the material should be sent to a testing laboratory—one that is approved by the federal Environmental Protection Agency. Technicians at the lab will analyze the material's fibers using a technique called polarizing light microscopy.

Most homes built or remodeled before 1978 contain some type of asbestos material (pipe insulation, for example). However, just because asbestos is present in construction material doesn't make it a health hazard. Asbestos is hazardous only when fibers are released into the air and inhaled. This happens when the material is damaged or in a condition known as friable—a point at which it can be pulverized by hand pressure.

Insulation for Ductwork

Our two-year-old house is fairly energy efficient. However, we have two large return ducts in the attic with only about 1 inch of insulation on them. I feel that I am losing a good deal of heat through them, but contractors and others don't think so. I think that

the air-conditioning ducts suffer similarly. What's your opinion?

I agree with you. Contractors were using 1 inch of insulation on ducts before the 1973 oil embargo. Since then, everyone has become conscious of the need to conserve energy. Attics that had previously been insulated with 3 inches of fiberglass for an R-value of R-11 are now being insulated with 6 to 9 inches of fiberglass (R-19 to R-30).

The more insulation on a duct, the less heat loss there will be in the winter and heat gain (for air-conditioning ducts) in the summer. However, there is a break-even point beyond which there are no economic benefits.

Because of the many variables involved—such as degree-days, the efficiency of the heating system, the number of hours the system operates, the velocity of the air moving within the ducts, and the cost per kilowatt-hour or per therm—you would need a computer to determine the optimum amount of insulation needed.

Nevertheless, the Small Homes Council–Building Research Council at the University of Illinois recommends a minimum overwrap around a duct of 3 inches of insulation. If the duct is to be used for air conditioning, it should be covered with a vapor barrier to prevent condensation within the insulation.

VENTILATION/MOISTURE

Attic Ventilation

I'm having a new roof put on my house. I have a roof-mounted fan that cools my house, and it's very efficient. The roofer wants to remove the roof fan and install a ridge vent. What should I do? Which of the two is best for my Cape Cod?

If your concern is cooling the house, then I would recommend the fan. I have been in thousands of attics over the years while doing home inspections, and those attics with roof-mounted fans (or power ventilators) were much cooler during the summer months than those with just a ridge vent with either gable or soffit vents.

A power ventilator is usually controlled by a thermostat, and it cools the house indirectly; that is, it cools the attic, which in turn reduces the heat load on the ceilings of the rooms below.

The attic must be adequately ventilated in the winter as well as in the summer months. However, a roof-mounted fan alone will not provide the necessary ventilation. You will still need gable vents or a combination of a ridge vent and soffit vents.

Roof Turbine Vents

My builder told me to cover the roof turbine vents during the winter to prevent warm air from being pulled out of the attic. This seems to make sense, so I stuff insulation in them before the cold weather arrives. The house seems to be warmer. What do you think?

I have seen many delaminated plywood roof decks that resulted from this type of thinking. The area directly below the roof must be adequately ventilated year-round to minimize moisture buildup. If you want to make the house more comfortable in hot and cold weather, install more insulation in the attic and seal cracks and spaces that let air leak from rooms below.

Normally, moisture rises into the attic because of cooking and bathing in the living area below, but it also can come from an uncovered dirt-floor crawlspace under the house. Excessive attic moisture also can lead to rotted lumber, damaged paint, and mildew.

Another benefit of a well-ventilated attic is that it keeps the underside of the roof deck cold, reducing the possibility of an ice dam developing. And in the summer, a well-ventilated attic reduces the ceiling temperature of the rooms below, and this reduces air-conditioning costs.

Covering Roof Turbines in Winter

I no longer cover my soffit vents during the winter since I read that they should be left open to permit air circulation. Does the same theory apply to revolving roof turbines? I don't know if it's better to leave them open as well, but I have seen many that are covered.

If the ventilation openings of the roof turbines are in excess of the attic's recommended free-air ventilation openings, then you can cover them. The recommended attic vent openings are 1/300 of the attic floor area when there is a vapor barrier on the underside of the insulation, and 1/150 of the attic floor area if there is no vapor barrier.

Many people cover the turbine vents in the winter for the same reason they cover gable vents—to keep attic heat in the attic. This can create condensation problems. The heat loss should be controlled by insulating the attic and not by blocking ventilation openings.

Adequate ventilation is necessary in summer to prevent heat buildup. In winter, however, ventilation is required to reduce or eliminate moisture buildup and condensation that can cause plywood roof sheathing to delaminate, paint to peel, and insulation to get wet and lose its effectiveness.

Attic Venting

The former owner of my house applied stucco on the exterior and ran the stucco under all the eaves out to the fascia board. The house has no eave vents. The only attic ventilators are two gable vents that are about 14

ATTIC VENTING

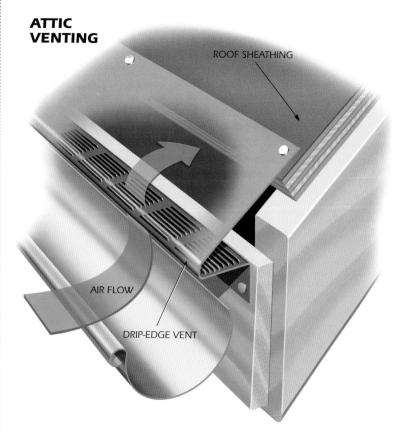

ROOF SHEATHING

AIR FLOW

DRIP-EDGE VENT

square feet. In other words, the gable vents are providing roughly half the amount of ventilation needed. Also, consider that the restriction of air flow, caused by louvers and insect screens, causes the effective free-air opening to be about 60 percent to 70 percent of the actual opening.

The best attic ventilation system is a combination of a ridge vent and continuous soffit vents. Rather than sealing off the gable vents and installing a ridge vent, you should install intermediate roof vents on both roof slopes and drip-edge vents at the eaves. These can be installed on existing roofs, without reroofing.

Mildew in Attic

I have mildew on the inside of my unheated walk-in attic. At one end of the attic, there is a crank-out louver window, and there are two roof vents on one side of the roof. I'd appreciate your reply.

Mildew on the attic side of the roof sheathing is caused by a moisture buildup on the surface, which in turn is the result of inadequate attic ventilation. Mildew, which shows up as dark spots on the roof sheathing, will not cause the roof deck to deteriorate. Once the attic is adequately vented, mildew stops growing.

The louver window should not be closed during the winter months—nor ever. Even if the totally unobstructed attic-ventilation opening satisfies the recommended formula of 1 square foot of opening for every 300 square feet of floor area, the distribution of the vents may not provide adequate air circulation. You will improve the overall attic ventilation if you install a gable louver at the other end of the attic.

inches wide × 20 inches long, one at each end of the 2400-square-foot attic.

What's the best way to provide attic ventilation in a case like this? I don't want to remove stucco.

Adequate attic ventilation is very important, and in your case drip-edge vents might be a way to provide it without cutting through or tearing off stucco. Some ventilation basics might be helpful here. Based on the area of your attic floor, the existing gable vents are incorrectly sized and will not provide a sufficient, unobstructed free-air opening. Assuming that there is a vapor barrier on the attic floor, the accepted minimum requirement for free-air ventilation openings is 1/300 of the attic floor area. Since your area is 2400 square feet, each gable vent should have an area of about 4 square feet, as opposed to the existing 1.9

Wet Attic

My two-story house is 40 feet wide on each side, and has three 12-inch-square vents on the roof's south side. There are four soffit vents on the house's north and south sides.

During the winter, frost collects on the attic side of the roof deck and on the rafters. The frost melts and drips on the insulation and seeps through the ceiling. What can I do to reduce the attic moisture in the winter?

Your problem is typical of an attic in the northern states that has excessive moisture buildup and inadequate ventilation. The unobstructed attic ventilation should be 1/300th of the attic floor area.

Based on your data, the vent openings are about 20 percent less than the recommended amount. If there are insect screens covering the vent openings, then the amount is even less. An insect screen reduces the effective opening by about 4 percent.

To increase moisture reduction, the roof deck between the rafters should be "washed" with cool, dry air. This can be achieved with continuous ridge and soffit vents. If these vents cannot be installed, then you must use additional roof and soffit vents. Frost tends to develop on the roof's north slope. There are no vents currently located there, so install the vents on the north side of the roof. Also, moisture can migrate into the attic through wall cavities because water can collect in the basement or crawlspace after a rain. Keep those areas dry.

Ice in Attic

I have a problem inside my unheated attic. During the winter, frost and ice form on the attic roof. I have 6 inches of fiberglass insulation on the attic floor. The attic has two windows, and if I leave the door leading to the

attic open, this seems to prevent frost and ice from forming. Is there a moisture problem? A new roof was installed about 12 years ago.

The frost buildup on the underside of the roof deck in the attic is not a moisture problem, it's a ventilation problem. Moisture vapor rises into the attic from living areas below. Cooking, showering and the ground below the house all release moisture into the house's air. Even with a vapor barrier under the attic insulation, some moisture will always find its way up to the attic because it leaks through open joints or seams in the vapor barrier, cavities in walls, or open sections around vent pipes or an interior chimney. The vapor enters the attic, condenses and then freezes on the roof deck. Over time, the moisture can damage the deck. In an adequately ventilated attic, the air movement will dissipate enough moisture so that frost will not develop on the cold surfaces.

You indicated that there are two windows in the attic. Are they closed? Quite often, attic windows are kept closed because they allow rain and snow to enter if left open. If the windows are open, are they located so that they provide cross-ventilation?

According to the Federal Housing Administration's minimum requirements for attic ventilation, the ratio of the total unobstructed ventilation opening to the attic floor area should not be less than 1:150, except when a vapor barrier is used on the attic floor, in which case the ratio may be 1:300.

If you are concerned that rain will enter the attic when the windows are open, you might consider replacing the windows with louvers or jalousie-type windows that deflect rain even when open.

Frosted Attic

The roof sheathing in our attic last winter was either heavily frosted or wet with large patches of mildew. The attic floor is covered with loose cellulose insulation but no vapor barrier. The gables are vented, and one end has a fan

that kicks in at 80°F. Do we need a plastic vapor barrier under the insulation and vents along the ridge? Our basement is damp, too. Does this contribute to the problem?

The problem in your attic is a reflection of the problem in your basement. A vapor barrier under the insulation will help reduce the frost and mildew buildup—however, it will not eliminate it because you have an excessive moisture condition in your house. Even with a vapor barrier, moisture will migrate up to the attic through plumbing and vent-pipe chases and other openings.

To control the attic problem, you will have to control the dampness in the basement. Make sure to clean your gutters and downspouts so that water can flow in the gutters and then down the downspouts. Also, the ground should slope away from the house, or surface water during a rain or snowmelt will flow toward the house and accumulate around the foundation.

If, after doing the above, there is still a moisture problem, you should coat the interior portions of the foundation wall with a cement-based sealer. I'm assuming you have a concrete floor in your basement, not a dirt floor. If not, install a concrete floor or cover the dirt floor with overlapped polyethylene sheets with the joints taped.

Mildewing Walls

After removing the wood shingles from the roof, filling in the roof deck spaces and re-covering the roof with asphalt shingles, we developed a mildew problem on the interior walls. Even though we installed two attic gable vents, the problem still exists. The house is not getting enough ventilation, but I'm not sure how to rectify this.

By filling in the roof deck spaces you altered the ventilation pattern of the house. The attic no longer breathes as easily as it used to. This causes a moisture buildup that results in the mildew.

Gable vents are helpful, but only if they are adequately sized. The effective openings for louvers is not the same as the actual opening cut out for those louvers.

The gross area of the actual opening can be reduced by as much as 65 percent because of the resistance introduced by the louvers. If you have a gable vent with outside dimensions of 2 x 2 feet, depending on the type of louvers and insect screen used, you may only have an equivalent opening of about 1x 1 foot.

An attic should have effective openings that are equivalent to 1/300 of the attic floor area. A single vent opening in an attic, even though it satisfies the total area requirement, is not considered adequate. The best method for ventilating an attic is to use a combination of vents such as soffit vents and a ridge vent.

Attic Moisture Problem

I have a problem with moisture condensation in my attic. Water drips from every nail in the roof, and it's not due to a roof leak. The attic has soffit vents, two windows, and a power ventilator with a humidistat installed in one gable. The ventilator runs constantly, adding about $20 a month to my electric bill. Can you provide me with some suggestions about how to solve this problem?

Your attic would be adequately ventilated under normal conditions, but you have an excessive amount of moisture finding its way into the space. In order to solve the problem, you must find the source of the moisture and deal with it.

The moisture source is probably your basement or a crawlspace. If either has a dirt floor, it should be covered with a plastic sheet. Joints between the sheets should be taped shut, as should the joint at the foundation wall. If the basement leaks after a rain, have the condition corrected by whatever means necessary. This may mean that cracks in the foundation wall and floor need to be repaired, a drainage system may need to be installed, as well as a sump pump. Gutters and downspouts should be in good working order. The water from gutters and downspouts should discharge far enough away from the house to prevent it from entering the foundation. Also, the ground should slope away from the house in all directions. If not, consider having the property regraded.

Hot Attic

My attic gets very hot during the day. There is an attic vent in the roof. How can I reduce the buildup of hot air in the daytime?

The vent openings in your attic may not be large enough to provide adequate ventilation. When there is a vapor barrier beneath the attic insulation, the unobstructed vent openings should not be less than 1/300 of the attic's floor area.

Even when an attic is adequately ventilated using passive vents, it will still be quite hot during the summer months. The temperature could easily build up to 140°F. The only way to reduce the high temperature is to exhaust the hot attic air and introduce cooler outside air. This can be done easily by installing a thermostatically controlled, roof-mounted power ventilator. Usually, the thermostat is set to turn the fan on when the attic temperature reaches 90° to 95°F.

I have inspected thousands of attics during the summer months, and have found that attics with power vents are considerably cooler than those with just passive vents.

Duct Condensation

In cold weather, water droplets fall from our air-conditioning ducts. The ducts are flexible plastic and lie on top of the attic insulation.

Solving the problem may be as simple as closing the dampers in the air-conditioning registers that are located in the rooms below the attic. During the winter, warm air may rise up and flow into the ducts. The ducts are surrounded by cold air and are usually uninsulated, so the moisture in the air condenses. Aside from closing the dampers, cover the ducts with batts of unfaced fiberglass insulation.

Attic Furnace and Ice Dams

My gas-fired, forced-air furnace is located in my attic. The heat from the furnace contributes to rapid snow melting on the roof above it, and this forms ice dams. Are there any solutions, such as installing soffit vents or insulation between the rafters?

According to the U.S. Army Cold Regions Research and Engineering Laboratory, the best way to prevent an ice dam from forming is to maintain a cold roof— that is, one where the temperature at the underside of the roof deck does not exceed 30°F. Maintaining a cold roof when the furnace is in the attic is difficult. The furnace, ducts, and metal chimney transfer considerable heat into the space. Insulating the ducts will help reduce the heat.

Passive vents alone will not generate sufficient air movement to dissipate the heat. You will need one or more motorized vents to exhaust the heat and draw in

ATTIC FANS

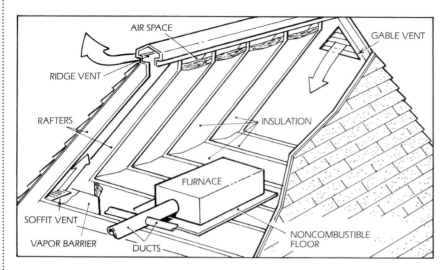

Diagram labels: AIR SPACE, GABLE VENT, RIDGE VENT, RAFTERS, INSULATION, FURNACE, SOFFIT VENT, VAPOR BARRIER, DUCTS, NONCOMBUSTIBLE FLOOR

cold air. Installing insulation between the rafters will help, but there must be an air space between the insulation and the roof deck. Finally, you should have a ridge vent and continuous soffit vents to provide air movement behind the insulation.

Attic Furnace Causes Leaks

My daughter has a problem with her new house. After a snow storm, ice builds up in the gutter and melted snow on the roof leaks through the asphalt shingles. She has a furnace in the attic with conventional duct insulation and a thick layer of insulation on the attic floor. Can anything be done to correct the problem?

The ice dam in the gutters is caused by the heat of the roof melting a layer of snow, which then refreezes when it reaches the gutter. Further melted snow is contained by the buildup of ice. This water backs up under the shingles and leaks through the roof.

The best way to prevent or minimize this problem is to keep the roof cold. Normally, insulation in an unfinished attic is located between the floor joists and not

between the rafters. This is fine as long as the attic space remains cold. In your case, the attic furnace is contributing to a heat buildup, which keeps the roof warm enough to melt the snow.

Insulating between the rafters with an appropriate air space between the roof and the insulation will help produce a colder roof. Ventilating this airspace is very important in reducing heat buildup and preventing condensation problems. This can be done with a ridge vent and continuous soffit vents.

To further reduce attic heat, cover the ducts with an insulation blanket. Don't forget that the attic area must be well vented, not only to reduce heat but to provide the necessary air for combustion in the furnace. I'd also recommend that you install at least one smoke detector in the attic as a safety precaution.

ATTIC FANS

Whole-house Fan

Since our home is surrounded by trees that shade the house and keep it cool, we are interested in using a whole-house fan rather than having central air conditioning. Please advise us on how this system works and where it can be purchased.

Although a whole-house fan is used for cooling, it does not actually cool the indoor air. Instead, it replaces the hot indoor air with fresh air from the outdoors. Even when the outside temperature is the same as the indoor temperature, cooling is achieved because air movement helps cool the body by evaporating perspiration. Air that's moving can feel as much as 10 degrees cooler than air of the same temperature that is still. Of course, a whole-house fan is most effective after sunset, when it can draw in cooler night air.

ATTIC VENTILATION OPTIONS

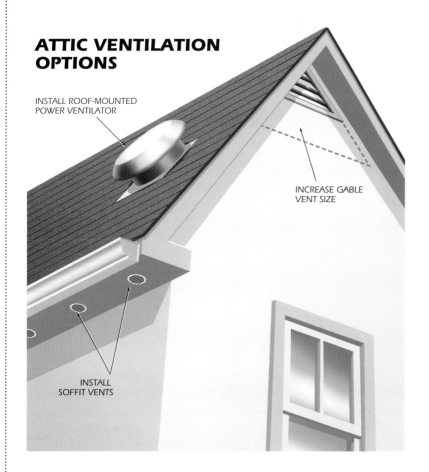

INSTALL ROOF-MOUNTED POWER VENTILATOR

INCREASE GABLE VENT SIZE

INSTALL SOFFIT VENTS

open when the fan is activated. This can cause condensation problems in the attic during the winter because the louvers, which are closed for months at a time, restrict air flow for ventilation. In order to prevent the attic from being inadequately ventilated during the winter, the hinged louvers should be held at a fixed, open position. I have seen a number of delaminated roof decks during my inspections that were caused, in part, by this.

Whole-house fans can be purchased at home centers. They often come with a manual on/off switch. However, the fan is more convenient to operate if this switch is replaced with a timer switch.

Getting Air Into and Out of the Attic

The fan is generally installed in the ceiling beneath the attic or in an opening cut in the gable end. When activated, the fan creates a negative pressure that causes warm indoor air to be pulled into the attic. Cool outside air is drawn in through one or more open windows.

When the fan is ceiling mounted, it causes the attic space to become pressurized, and this causes the hot air to be pushed out of ventilating louvers such as gable or soffit vents. When the fan is gable mounted, it pulls the air out of the attic. In either case, it's critical that one or more windows be kept open, not only for cooling but for safety reasons. Unless adequate makeup air is supplied, the negative pressure created in the house can pull in exhaust gases from combustion appliances, such as a gas-fired water heater. This is obviously dangerous.

Gable-mounted fans have hinged louvers facing the outside that are normally closed when the fan is off and

I have a brick house with a gable roof. It has very little attic space. I have looked at a lot of similar homes and none have a rooftop (attic) fan. Is it possible that my attic is too small to install an automatic rooftop fan? Any information would be helpful.

I doubt that your attic is too small for an attic fan, since the fan does not project into the attic space. It is mounted above the roof surface, where it is covered with a dome to protect it from the weather. The main concern when installing a fan is the need to provide an adequate amount of intake air so that the fan does not create negative pressure in the house. This would cause a natural draft gas-burning appliance, such as a water heater, to backdraft.

For air intake vents, attic-fan manufacturers recommend an unobstructed opening of 1 square foot per

300 cubic feet per minute (cfm) of fan capacity. If you have gable vents, you may have to increase their size to accommodate an attic fan while allowing for the fact that the louvers on the gable vents decrease the vents' effective opening by as much as 50 percent. However, the most effective location for the air intake vents is on the soffit. The cfm requirements for an attic will depend not only on the area of the attic floor but the pitch of the roof and whether the roof covering is a light or dark color.

Attic Heat

I have a three-year-old, chalet-style house with about 2400 square feet of living space. About 1000 square feet of that space is attic. The rest is divided between floor-level and loft-living space. The attic gets extremely hot in the summertime. What's the best way to reduce the heat in the attic? Should I use a ridge vent, an attic fan, or a roof-mounted ventilator? Any help you could give would be appreciated.

The two items of concern regarding attic ventilation are moisture control in the winter and heat removal in the summer. Without ventilation, an attic's summertime temperature can easily build to over 140°F. Since removing heat requires more air movement than removing moisture, the venting system should be designed for summer conditions.

From an energy-conservation point of view, a ridge vent with continuous soffit vents is the most efficient since it relies only on air convection. However, if your main concern is to reduce the heat buildup, then you might try a thermostatically controlled roof-mounted ventilator. The thermostat is set to the temperature at which you want the fan to start, usually around 95°F. Although I haven't done a scientific survey, my experience from inspecting thousands of attics during the summer is that attics with power vents were not as hot as those with ridge vents.

TRUSS UPLIFT

Roof-truss Uplift

Every winter, the ceilings in two of our upstairs bedrooms separate from where they meet the walls by $\frac{1}{4}$ inch or more. In the spring, the cracks close up. What causes this? Is there a way to prevent it?

The condition you describe is roof-truss uplift. It's caused by swelling and shrinking between the bottom and the upper chords of the roof truss. The bottom chords are covered with insulation, while the upper chords are exposed to cooler air in the attic. In the winter, the top chords can gain moisture from condensation on the roof sheathing. This causes them to swell in length. Meanwhile, the bottom chords are exposed to high temperatures and a low relative humidity. They can shrink slightly. This causes the upper chords to pull the bottom members up, lifting the ceiling below.

A simple cosmetic solution is to fasten molding to the ceiling. The molding will move with the ceiling and keep the crack covered.

To avoid shifting damage to rooms below the attic, steps are usually taken during construction to allow truss movement. For example, there is a bracket that

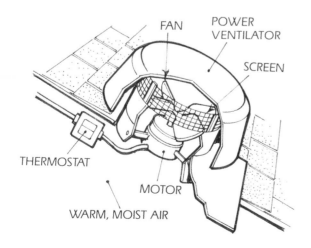

FAN
POWER VENTILATOR
SCREEN
THERMOSTAT
MOTOR
WARM, MOIST AIR

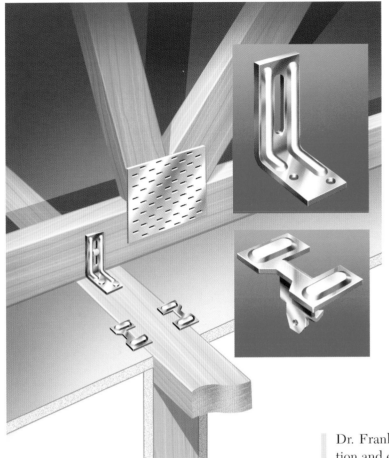

More on Truss Uplift

Earlier, we discussed roof-truss uplift, a condition in which the joints between the ceiling and walls crack and open. The condition occurs in the winter and is caused in part by differential shrinkage between the bottom members (chords) of the truss, which are covered with insulation, and the upper truss members, which are exposed to the cooler attic. The bottom chords are exposed to high temperatures and a low relative humidity, and they shrink and shorten slightly compared to the upper members. This causes the upper chords to pull the bottom members up, lifting the ceiling below.

connects the truss to the wall partitions below but allows the truss to move (refer to illustration).

The bottom of the bracket is nailed into the wall plate, and another nail is driven through the slot into the truss. As the truss moves, the nail slides up and down in the slot. Incidentally, this isn't to say that hardware can make up for the humidity caused by inadequate attic ventilation. The attic should be properly ventilated and insulated, and air gaps that allow moisture and heat to pass into the attic should be sealed. The steps taken to allow for truss movement are the extra insurance that keep minor fluctuations in the truss from causing damage to the ceiling below.

Dr. Frank Woeste, P.E., professor of wood construction and engineering at Virginia Polytechnic Institute, was kind enough to send us additional information on the subject. According to Dr. Woeste, the greatest impact on truss uplift is top chord swelling. In the winter, because of condensation on roof sheathing, the top chord can gain moisture and swell in length. This action is additive to the bottom chord shrinking, and causes additional truss uplift.

Truss uplift has been recognized as a construction problem for at least 30 years. Researchers examining the problem found that it is reduced by lowering the relative humidity in the attic. Uplift, they say, is more common in homes where the lower truss chords are covered with thick insulation and where moisture from bathroom and kitchen exhaust fans and clothes-dryer vents discharge into the attic. Moisture should be vented outside the house.

MISCELLANEOUS

Hidden Attic Stairs

The ceiling-mounted hatch for my folding attic stairs sags down about 1½ inches. How do I correct this problem?

During my years of doing home inspections, I have seen this many times. It's a common problem that should be corrected for two reasons: It looks ugly, and the gap is a source of considerable heat loss.

I contacted two manufacturers of folding attic stairs, and both said the springs that pull the stairs up are adjustable. There are two or three holes to which the springs could be attached that will increase the springs' tension. If that doesn't correct the problem, you will have to replace the springs. Call the manufacturer. There should be a name and model number on a side rail or under a step.

Note that the folding stair springs are under tension, so never adjust or remove them when the stairs are open.

Emergency Escape

I'm thinking of converting my third-floor attic area into a bedroom. Is there anything I should be particularly concerned about?

Your main concern should be to provide a safe exit in case of fire. Usually, there is only one stairway to the attic. If a fire develops in the floor below, the stairway may not be safe to use. If there is no exterior fire escape from the attic, then a rope or folding ladder should be provided. Several manufacturers offer UL-approved ladders designed for this.

Also, the window opening must be large enough to permit easy escape. Some municipalities require a minimum open height or width of 20 inches and a total minimum open area of 5.7 square feet, and the bottom of the opening must be no more than 44 inches above the finished floor.

Check with your municipal building department for the specific regulations in your area. Before beginning construction, you or your contractor must get a building permit. And, you should adhere to all applicable building and electrical codes so you can obtain a Certificate of Occupancy (CO) for the new room.

Attic Load

Three years ago, I acquired an English-style 1½-story, all-brick, all-plaster house, 50 years old and in excellent condition. There is a full attic with a catwalk in the center. I would like to finish the attic, but it only has 2 x 6-inch ceiling joists, 16 inches on center. I am told the joists have to be at least 2 x 8 inches, 16 inches on center before rooms can be added. First, why were houses like this built, and what's the remedy? Can I do most of the work myself?

When designing a house, the size of the floor joists depends on the distance between the joist supports (span), the on-center distance between the joists (usually 16-inches on center), the species of wood the joists are made of, and the loading applied to the floor. In residential structures, the floor loading is usually designed for a live load of 40 pounds per square foot, with attic floors designed for loading of 20 pounds per square foot. A live load is the load that is imposed solely by occupancy (persons, furnishings, and appliances).

The attic's design load is less than the load for habitable rooms because the attic is normally not used for storing heavy appliances or furniture. Hence, ceiling joists are often smaller than the floor joists below.

In determining joist size for a given span, architects use tables that consider bending stress, deflection, and stiffness (in order to minimize the springiness in the floor). For a rough estimate of joist size and span, you can consult one of these charts. Your floor joists are sufficient for a 40-pound-per-square-foot floor, providing their unsupported span does not exceed 9 feet 6 inches (approximately).

To finish the attic properly, you would need more than such a chart, however. In order to receive a building permit to begin such a project, your local municipal building department will probably require that you have plans drawn by a registered architect or approved by a licensed professional engineer.

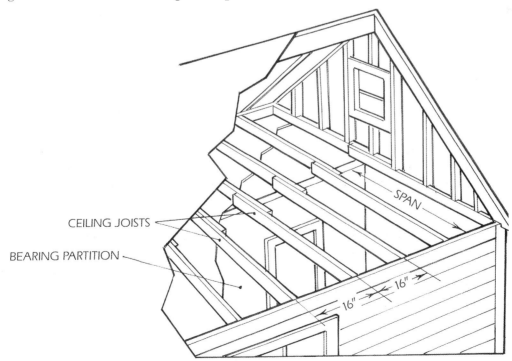

CEILING JOISTS

BEARING PARTITION

SPAN

16"   16"

9. Interior Rooms

CEILINGS • FLOORS • WALLS • WINDOWS • FIREPLACE • BATHROOM • KITCHEN • LAUNDRY ROOM • DAMPNESS, HUMIDITY • FIRE SAFETY • CEILING FANS • MISCELLANEOUS

CEILINGS

Joist Shadow Lines

There are dark lines on the drywall ceiling below the attic in my house. The lines coincide with the ceiling joists. Someone suggested it is because of a lack of insulation. However, we have 6-inch batts in the attic. Do you have a cure for this problem?

Although you have insulation between the joists in the attic floor, the tops of the joists are exposed to the low winter temperatures in the attic. Since the wood joists are not effective insulators, they act as thermal bridges. Consequently, the temperature at the underside of the joists (at the drywall ceiling) is lower than the adjacent sections of the ceiling that are covered with the insulation batts. Because of the lower temperature below the joists, condensation (however slight) tends to form along these areas. Over time, the moisture traps dust and also results in mildew growth, which shows up as shadow lines.

To prevent this from recurring, first paint the ceiling. Use a paint containing mildewcide. Next, install insulation batts over the exposed ceiling joists. Ideally, the insulation should fill the spaces between the joists and cover the tops of the joists as well. This last layer of batts is installed perpendicular to the joists.

However you install the insulation, make sure to use a type that does not have a foil or kraft-paper vapor barrier. And be sure additional insulation does not cover soffit vents or recessed light housings (unless the housings are IC types rated for direct contact with insulation).

Bulge in Plaster Ceiling

I live in an old house with plaster ceilings. When I moved in, I noticed that the ceiling in one room had a bulging section. Is this something I should be concerned about?

The sag may well be something to be concerned about. Older homes often have plastered ceilings and walls. The plaster is applied over wood, metal, or gypsum lath. The first plaster coat is applied with just enough force so it squeezes between the spaces in the lath and into the wall cavity behind it. The plaster slumps over the lath as it squeezes past it and hardens into keys, locking on the first coat.

The first layer is scratched with a trowel so the next coat can adhere to it. The next coat is called the brown coat and the final one, the finish coat.

Over time, vibrations and wetting from roof and plumbing leaks may weaken the keys, cracking them. The combined weight of the three coats causes the plaster to break away from the keys, forming the sag you mentioned. In cases where the house has been unheated for prolonged periods, the finish coat alone may separate and sag, though the keys are intact.

You can call in a plaster contractor (not the same as a drywall contractor) and have the sag removed and replastered, or you can try to repair the area with plaster washers.

Drive a 1¼- or 1½ inch No. 6 drywall screw through the washer's hole and into the ceiling framing above the plaster. The washer supports the plaster around the screw as the screw pulls the plaster against the framing. Driving the screw into the framing is more secure than driving it into the lath, in which the screw may or may not grip. Encircle the sag, driving the screwed-in washers 6 to 10 inches apart. Once the out-side perimeter is secured, move inside the perimeter and repeat the process in concentric circles until the sag is fully secured. The washers are perforated to hold the Spackle that covers them.

Dark Areas on Walls/Ceilings

We have been getting black soot-like marks on the walls and ceilings in our 6-year-old home. It seems to cling to where the studs are in the walls and also by the outside corners. What causes this, and can it be prevented?

Depending on the tightness of your house, and the extent to which there is a negative pressure in it, the black areas on your walls and ceiling could be soot from burning candles or from other sources of combustion, such as a pilot light that is out of adjustment.

However, based on the location of the black areas, I believe they are the result of mildew and dust. The condition is quite common and is caused by a thermal bridge. Even though there is insulation in the exterior walls, the batts are separated by studs, which are poor insulators by comparison. As a result, the portions of the wall fastened to the studs will be cooler than the other areas, because more heat is lost through the studs than through the insulated portions of the wall. The same holds true for the outside corners where the walls meet. The wood framing at the corners has very little room for insulation.

Because of the lower surface temperature by the wall studs and corners, a small amount of condensation tends to form on these areas. Over time, the moisture traps dust and provides conditions for the growth of mold and mildew.

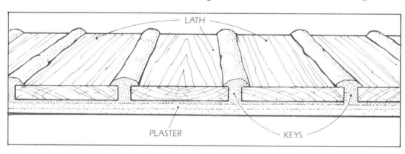

LATH · PLASTER · KEYS

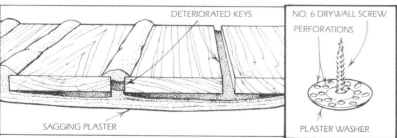

DETERIORATED KEYS · SAGGING PLASTER · NO. 6 DRYWALL SCREW · PERFORATIONS · PLASTER WASHER

Periodic cleaning with a solution of bleach will kill the mold and mildew. Eventually, the walls should be painted with an interior paint containing a mildewcide.

Ceiling Stains

We have a moldy ceiling stain in the hall. The leak has since been repaired, but we are wondering what we can do to clean up the stain.

If the ceiling only has a mold buildup, you can clean it with a solution of water and bleach. A powerful solution consists of three parts water to one part bleach. Dab it on the ceiling, and wait 20 minutes before rinsing it off. Blot the area dry with a paper towel. However, once you remove the mold, you may find that a pale brown stain remains on the ceiling. You can't simply paint over the stain. After a few months, the stain will bleed through the paint. To be sure that the stain does not resurface, apply a primer sealer.

Water Spots on Ceiling

I was wondering if you have a solution to my problem. A leaky roof left water spots on my bedroom ceiling. I have had the roof fixed and I have tried to paint over the spots, but they just show through.

There are a number of products on the market that will solve your problem. They are available at hardware stores, paint stores, and home centers, and are referred to as water-stain blockers or sealers (also called stain-blocking primers and stain-blocking sealers). The stain-blocking sealer/primer is applied to the ceiling with a brush or a roller, or it is sprayed on (it's available in aerosol cans). After the sealer/primer is dry, paint the ceiling.

Asbestos in Ceiling

My house has a sprayed-on "popcorn" ceiling. I've read that such coatings contain asbestos. Is this true and is it a health hazard?

The only sure way to know if your "popcorn" ceiling contains asbestos is to have a sample analyzed by a testing laboratory. If it does contain asbestos don't panic. Simply containing asbestos doesn't make it a health hazard. If the material is in good condition and is unlikely to be disturbed, then any effects of the asbestos are considered negligible.

If your ceiling contains asbestos and you have deteriorating or damaged sections, asbestos fibers may be released into the air, creating a health hazard. In this case, contact an asbestos abatement company and have the ceiling removed.

Do not simply scrape the "popcorn" off because you'll be releasing asbestos fibers. Contact your local health department or the Environmental Protection Agency for the names of licensed abatement companies.

Spots on Ceiling

The painted ceiling in my bathroom has developed brown spots. I know that the spots are not caused by a water leak from the room above and they are not caused by mildew. Do you know what they are?

Latex paint contains a number of components, many of which are water soluble. When a surface coated with latex paint is exposed to very high humidity, the water-soluble components in the paint tend to leach out and appear as brown spots. These can usually be removed by scrubbing them with a water-dampened sponge. You don't need an abrasive cleaner. It may take a couple of applications with the sponge, but, once the spots are removed, they shouldn't come back.

FLOORS

Terrazzo Care

What are the correct procedures to care for terrazzo tile? Should it be waxed, buffed, stripped yearly and rewashed again? Some say never wax or strip terrazzo tiles, others say the opposite.

For readers unfamiliar with terrazzo, it is a concrete floor with marble chips embedded in it. The chips must cover 70 percent of the floor area. The floor can be either cast in place or precast and ground smooth. The other 30 percent of the floor consists of a portland cement or epoxy binder.

Once a year, the floor should be stripped of old sealers and resealed with a product designed for terrazzo. A terrazzo floor should never be waxed with a general-purpose floor wax, as this results in a discolored and slippery floor.

The National Terrazzo and Mosaic Association (NTMA) recommends using pH-neutral detergents that have been formulated for terrazzo use. The association says that the cumulative effect of using soaps, rather than detergents, will cause the floor to become dull and lifeless.

The NTMA recommends sweeping the floor daily using a yarn-wick brush treated with nonoily sweeping compound (available from janitorial supply houses). Scuff marks and stains should be removed by hand with a neutral cleaner diluted in warm water.

Lightly soiled floors should be damp mopped weekly, using a neutral cleaner. Heavily soiled floors should be scrubbed with a mechanical buffer and neutral cleaner.

Covering Terrazzo

I am about to remodel my 30-year-old home, and I would like to cover its terrazzo floors with carpet or vinyl. My floor has a small number of cracks, none of which allow ground water to enter. How do I prepare the terrazzo for the new flooring?

If you cover the floor with carpet, no special treatment is needed. However, you will have difficulty securing carpet tack strips at the corners where the floor meets the walls. In most cases, the strips are nailed down, but in your case they will need to be attached with adhesive because terrazzo is extremely hard—it consists of concrete with polished marble chips embedded in it.

If you cover the floor with vinyl, first remove any wax or sealer by washing the floor with a solution of two cups of ammonia in a gallon of water. Similarly, all cracks should be cleaned and filled with epoxy grout. After the floor has been prepared, adhesive can be spread on it and the vinyl can be installed.

Tiling Radiant-Heat Floor

I have a home with radiant heating (cement-slab floor with embedded circulating hot water), and I want to replace the existing vinyl tile in the kitchen. I would like to use slate, quarry tile, or something similar, but I'm concerned about the disruption of heat transfer and the chance of increased energy use. What do you recommend as the best covering for this type of floor?

Any floor covering that you like will do fine. I checked with the American Society of Heating, Refrigerating and Air-Conditioning Engineers (ASHRAE) and they say there would be no appreciable increase in energy use regardless of the type of floor covering.

The response time of a radiant-heating system to changes in thermostat setting is normally slower than that of a hot-water system with baseboard or free-standing radiators, or a forced-warm-air system. This means it takes longer for a radiant system to reach the desired comfort temperature than the other two systems. By covering the floor, you will slow down this response time slightly.

Hardwood Flooring over Concrete

I would like to install a strip oak tongue-and-groove floor on top of an above-grade concrete slab. However, I've heard horror stories of costly wood floors buckling from moisture. How can I avoid problems caused by moisture that might accumulate in the area under the slab? I'd appreciate your advice.

A hardwood floor can be installed on a concrete slab at or above grade. The Oak Flooring Institute recommends against below-grade installations.

Moisture is the chief culprit in hardwood-floor buckling, so test the slab for dryness. Tape 1 square foot of clear polyethylene sheet to the slab, sealing its edges with plastic tape. If, after 24 hours, no clouding or moisture droplets have formed under it, the slab is dry enough for you to install a wood floor over it.

To prevent moisture from reaching the underside of the hardwood floor, place a vapor barrier of either building felt or polyethylene plastic over the slab prior to the installation, as shown in the drawing.

Wood Floor Over Concrete

There is a 1¼-inch-wide x 24-foot-long crack in the concrete floor slab of my 31-year-old home. We live in San Diego County so moisture is not a big issue. We want to lay a hardwood floor over the slab, but we are not sure how to proceed.

You didn't mention whether the slab is below grade or not. In most cases, installing a hardwood floor over a below-grade slab is not recommended. When the slab is below grade, there is always the risk of moisture related problems resulting from surface and subsurface water seepage. Even if moisture is not a big issue, at the very least you will need a vapor barrier, such as

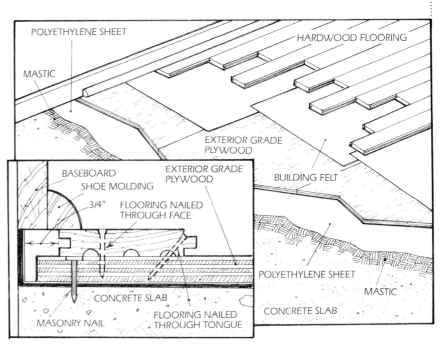

6-mil polyethylene sheets, between the concrete slab and the underside of the wood floor.

The hardwood flooring strips or planks should not be mechanically fastened to the concrete slab. They should be fastened to an approved subfloor such as ³/₄-inch or thicker sheathing-grade exterior-use plywood. The subfloor can be fastened to the concrete slab with either concrete nails or glue. It also can be supported on screeds (sometimes called sleepers).

Resilient Tile Over Concrete

I have a 1700-square-foot house with ceramic tile over concrete. Can the tile be removed and replaced with resilient tile or sheet vinyl?

Any type of vinyl flooring can be installed over a concrete floor as long as the area is smooth, clean, and dry. The big question is whether the floor is dry. Some amount of moisture will normally be emitted from the concrete slab, but excessive moisture will cause the new tiles or sheet vinyl to lose their bond to the concrete. In turn, this causes tiles to become loose or seams to curl in sheet vinyl.

A flooring installer needs to check the amount of water vapor that is passing through the concrete before installing any flooring. Although homeowners can get an idea of whether the floor is dry by taping small squares of plastic on the floor and then checking them for condensation, professionals can make a more accurate estimate. They can actually measure the amount of vapor (expressed as the weight of the water measured in pounds per 1000 square feet of floor) emitted over a 24-hour period.

Manufacturers of vinyl flooring make recommendations for the amount of moisture vapor that their floor system can tolerate. For example, to install vinyl with felt backing, the floor must emit no more water vapor than 5 pounds per 1000 square feet of floor in 24 hours.

Lifting Resilient Tiles

The floor tiles in my 11-year-old home seem to be lifting. When we step on them, they make a crunching sound, as if there were sand under them. What causes this?

I assume that you are referring to tiles on a concrete slab. If this is indeed the case, moisture is collecting beneath the tiles and causing them to lift. The crunching sound is probably efflorescence under the tiles. Efflorescence is caused by soluble salts in the concrete slab that dissolve in the water as it migrates through the slab. When the water evaporates, the salts remain encrusted under the tiles. This condition can occur even though you do not see standing water on the floor. To prevent the problem, you will have to reduce or eliminate water below the slab with a drainage system around the perimeter of the foundation. However, installing such a draining system is both expensive and disruptive to the house's landscaping. You would have to decide whether it would be better to install drainage, use another type of flooring, such as carpeting, or simply replace lifting floor tiles periodically.

Removing Linoleum

We'd like to refinish the hardwood floor that's under the linoleum in our kitchen. Is there an easy way to remove the linoleum?

Unfortunately, there is no easy way. If the hardwood floor consists of oak strips and was originally smooth, then the linoleum was probably glued directly to the floor. Begin by cutting the linoleum into 12-inch strips with a utility knife. Be careful to set the blade depth so that you don't cut into the wood. Use a long-edged trowel to pry up the linoleum strips. A heat gun will help to soften the adhesive as you go. Eventually, you'll remove all the linoleum and some of the adhesive. Portions of the linoleum's felt backing, however, will stick to the adhesive that remains.

If your hardwood floor had open joints or crevices, it was probably leveled with a quick-setting plaster-like compound before the linoleum was applied. In this case, prior to lifting the strips, pound on the linoleum with a flat object to crumble the leveling compound and facilitate lifting.

After removing the linoleum, you'll have to sand the floor. Be aware that it is possible that the felt backing of linoleum contained asbestos. You can check this by having a sample analyzed or contacting the manufacturer. If it does contain asbestos, and the floor is sanded, then asbestos fibers will become airborne, creating a health hazard. In this case, you'd be better off simply covering your floor with a new layer of linoleum.

Covering Old Vinyl Flooring

The house I bought has two bathrooms, and I want to put new vinyl flooring in them. I'm not sure if the old vinyl has asbestos, or even whether that matters, but in any case I would like to know how to go over the top of the old vinyl.

As long as you cover the existing vinyl, it doesn't matter if it contains asbestos. However, manufacturers don't recommend covering over more than one layer of vinyl flooring. In that case, manufacturers recommend removal. Then the presence of asbestos becomes a concern. Unless you're certain that the vinyl flooring does not contain asbestos, for safety's sake you should assume that it does.

Asbestos in vinyl products is non-friable (that is, when it is dry, it cannot be crumbled, pulverized, or reduced to powder by hand pressure). But some removal practices, such as grinding or sanding, can cause the particles to become airborne.

If the existing vinyl flooring consists of a single layer, you can cover over it as long as it is well bonded. The vinyl should be cleaned, and any surface finish such as wax must be removed by wet stripping only. If the vinyl is textured or embossed, you should use an embossing leveler to fill in and smooth out the surface. Another option is to cover the floor with $\frac{1}{4}$-inch-thick plywood, and then apply the flooring.

Asbestos Floor Tiles

I am purchasing a house with hardwood floors. Unfortunately, two of the bedroom floors are covered with 8-inch asbestos tiles. I've been advised to leave them there and put another type of flooring material over them. I'd like to have the tiles removed. Is it best to leave them or can they be removed to reveal a nice floor?

If you remove the tiles, you may be disappointed to learn that the floor under the tiles is not hardwood. Just because there are hardwood floors elsewhere in the house doesn't mean that all the floors are hardwood. I have seen homes where the perimeter of the floor around an area rug was hardwood, and the section under the rug was plywood.

According to the Resilient Floor Covering Institute, the removal of resilient floor tiles should be the last alternative, after all the others are considered. Not only is there a concern that the tiles contain asbestos but that the adhesive used to secure the tiles may also contain asbestos.

Removing asbestos floor tiles is not an easy task, and if that is your decision, it's best left to a contractor that observes federal guidelines for occupational exposure to asbestos.

Squeaky Floors

My six-year-old house has carpeted floors that squeak. The best solution I can think of is to remove the carpet, then go into the crawlspace and somehow force construction adhesive between the joist and the subfloor, then screw through the floor from above and into the joist. Is there an easier or better solution?

First, I'm assuming that you don't have hardwood flooring under the carpet because individual hardwood flooring strips can squeak.

If you don't have hardwood floors below the carpet, your solution will work, but because of the crawlspace it will be very difficult. You may consider first driving screws from above, to see if that prevents the squeaking. If the screws alone don't do the job, you can always back the screws out and try your method.

Another option you should consider, before resorting to more difficult repair methods, is using a proprietary tool system called Squeak No More. With it, you drive a screw right through the carpet into the joist and then snap off the screw head using a tool designed for that purpose. The screw is specially made. Its shank is scored so that it snaps off inside the subfloor. The product costs about $20 at home centers.

I built a stilt house that is 15 feet off the ground. I put down a ¾-inch-thick tongue-and-groove plywood floor over 2 × 12 floor joists and covered it with carpeting. The floor now squeaks in several places. How can I fix it?

The type of floor you have is called a single-floor system. It was developed by the plywood industry and can be used in place of the standard floor, which consists of two panels (subfloor and underlayment).

The squeaking is caused by the floor panels rubbing against one another or against the framing. It's a fairly common condition and usually results from the loosening of some of the nails holding down the subfloor

because of shrinkage or warpage of the framing. This generally occurs when smooth-shank nails are used to fasten the subfloor rather than ring-shank or spiral-shank nails, which are less likely to pull loose. It's also possible that not enough nails were used to fasten the floor to the joists. In that case, the squeaking should be reduced by renailing the loose sections.

According to the American Plywood Association, you should use 6d ring-shank or spiral-thread nails. Also, the nails should be spaced 6 inches on center at the panels' edges (where panels butt each other), and 10 inches on center at intermediate joists.

Squeaking Hardwood Floors

I would appreciate your suggestions on stopping squeaks in hardwood floors. Mine are located in the second-floor bedrooms. These floors are not accessible from below, unless you tear out ceilings in the kitchen, living room, and dining room. Is there a way to stop the squeak without ripping down ceilings?

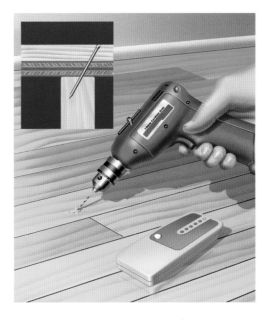

Hardwood floors squeak when someone walks over the floor strips and causes them to rub together or rub on a nail. Another source of the noise is the subfloor moving against the floor joists. You can tell the difference. If the squeak is in one or two strips, it's likely that they are rubbing together. If the floor squeaks when you walk over an area that is 1 to 2 feet wide, the cause is probably subfloor movement.

Correcting a squeak caused by movement in the subfloor will require you to reattach the subfloor to the joists. This is best done by exposing the floor joists from below and screwing through the joist into the subfloor, but the resulting mess and repairs to the ceiling make this unattractive in most situations.

An alternative is to fasten the subfloor to the joists from above. This requires that you drive 2½-inch-long finish nails through the finish floor into the joist. Locating the joists is difficult without a stud finder, and not all of these devices can find a joist through hardwood flooring. The cost of a stud finder that is designed for difficult sensing will easily repay itself in making this difficult job less so. The device will also prove itself handy in other applications.

Once you locate a joist, bore a pilot hole at an angle through the hardwood floor, the subfloor, and into the joist. Drive and set a finish nail and putty over it.

When the squeak is caused by movement in the floor itself, the Oak Flooring Institute recommends that you squeeze some liquid wax, talcum powder, or powdered graphite between adjacent floor strips where the noise occurs. If that doesn't work, try driving triangular glazing points between the strips. Angled-face nailing with finish nails may stop the squeak, but it's not necessary to drive the nail into the joist.

Wood Floor Care

The floors in my home have always been cleaned and waxed with an electric polisher. However, wax buildup makes the floors look as though they need to be cleaned. What solvent safely removes wax buildup?

Paint thinner (also known as mineral spirits) is sometimes used to remove wax buildup from floors. It's flammable, however, so it must be handled with care. Open windows and doors to keep the area well ventilated. You should also wear rubber gloves and put on a respirator that's equipped with filter cartridges for organic vapors. Paint thinner, the respirator, and the gloves are all commonly available at hardware stores, paint stores, and home centers.

Warped Wood Flooring

For the first time since it was installed 10 years ago, our tongue-and-groove floor has buckled. Moisture appears to have gotten under it. When it dries out, will it return to normal, or should the area be replaced?

First, find the moisture source and prevent it from reaching the floor again. Then let the floor dry out over the course of a heating season. When it dries out, it should return to a relatively flat position. However, you may notice discoloration, looseness, and some cupping even after the floor has dried. Unless the damage is severe, replacing the damaged area should not be necessary. After loose sections are refastened and the area is sanded, the entire floor should be refinished to blend in the repair.

Flexing Floor

In the home I built I used 2x8 floor joists spaced on 16-inch centers and spanning 12 feet. The subfloor is nailed down with spiral nails and the finish flooring is glued and screwed. Over the 12-foot span there is one row of metal bracing. My problem is that the floor springs when I walk across it. How can I strengthen the joists?

According to span tables, for a 40-pound-per-square-foot live load (typical design criteria for residential floor joists), your floor should not show excessive deflection or bounce. However, the tables are based on specific grades of wood such as Douglas fir and hem-fir. Since you didn't mention the grade of wood you used, I assume that its bending stress falls within the acceptable limits; however, its stiffness factor may be marginal. You can usually eliminate springiness by nailing "sister" joists of the same size to every other existing joist. When installing these joists, you may need to notch or shave the ends so they fit between the subfloor and sill. Although I'm sure this won't be necessary, you can always add more sister joists to the remaining joists to stiffen the floor further.

Slab-house Problem

My husband and I live in a two-story house in which the first floor is on a concrete slab. Is it healthy for us to live on the slab? Can gases come through the carpet and pad from the slab? When we had new carpet and pad installed on the slab the installers discovered two or three large cracks, which they patched. Is this safe?

The only potential problem that I am aware of is radon gas seeping into the house through cracks, open joints or porous sections in the concrete slab. However, sealing cracks in the slab, as was done, is a good first step in minimizing this. In order to determine whether the house's radon concentration is above the level recommended by the Environmental Protection Agency, you should run a long-term test. If the test shows elevated levels of radon, you should contact a radon-mitigation contractor listed with the EPA.

Floor Covering, Before or After Painting?

Please settle a repeated difference of opinion. When someone's rooms need new carpets or tile, is it better to put down floor covering before or after you paint walls and ceiling? Some floor-covering people say put down new covering and then paint. I don't agree. Who wants dirt on new floors? What do you say?

There is no definitive answer. It is purely a matter of opinion, and the quality of care and workmanship on the part of the painter and installer. If the painter and carpet/tile installer are meticulous workers, then it doesn't matter whether the rooms are painted first or the floors covered first. If the workers are not careful in attending to details, then in my opinion the rooms should be painted first and then the floors covered. My only concern is the scuffing of the baseboard when the carpet is being stretched. However, it is easier to touch up or paint a scuffed area than to remove paint spatter from a carpet.

Carpet Indentations

How can depressions in carpets caused by tables, chairs, and other furniture be removed?

I've had success removing carpet indentations with a steam iron. To do this, you hold an iron over the depression and apply steam to it. Don't press the iron down on the spot. Saturate the area with steam, then rake the fibers with your fingers. Depending on how old and deep the depression is, it may be necessary to steam the area several times to remove it.

Static Shock

About four months ago I had a new carpet installed. Recently, I have been getting shocks when I walk on the carpet and then touch the stove, sink, light switch, or anything that is grounded. Can you tell me how I can fix this problem?

The problem you're experiencing is the discharging of static electricity. Rubbing or moving two dissimilar nonconducting materials against one another brings about this condition. It is usually more of a problem in the winter when the relative humidity is very low, than in the summer when the humidity is generally high. This is a clue as to how you can minimize the problem. Studies have shown that when the relative humidity is maintained in the 40 to 50 percent range, the conditions that produce static electricity are greatly reduced or eliminated. Moist air is a better conductor than dry air and, as such, helps dissipate the change by providing a path for the static electricity to flow to ground. Increasing the humidity in your house is simple enough if you have a warm-air heating system. Just install a furnace humidifier in the ductwork. Otherwise, you can use one or more stand-alone humidifiers to add moisture to the air.

WALLS

Finding Wall Framing

I've been doing a lot of remodeling, and discovered a quick way to find wall studs. Drill a tiny hole on each side of a wall outlet to find the stud the electrical box is mounted to. The next stud should be 16 inches away in either direction.

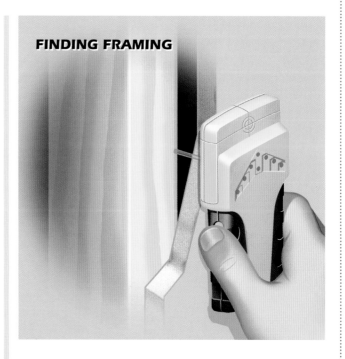

FINDING FRAMING

Your method will work most of the time, but wall receptacles are not always adjacent to a stud, and many houses have studs spaced 24 inches on center. That's why an inexpensive stud sensor is invaluable. When you use a stud sensor, all you have to do is run it across the wall surface. When it passes over a stud, it emits an audio or visual signal—or both.

Grab Bar

I had a company out to install a bathtub grab bar, but they could not successfully locate the wall studs. They recommended some other methods of installing a grab bar instead of attaching it to studs. What method do you recommend?

The only method I recommend for securing a grab bar is to fasten it to the studs. A person can exert considerable force on a grab bar, and if the bar is not adequately fastened, it will probably pull out of the wall, rather than being loosened.

Many people mistake the washcloth rail on a ceramic soap dish for a grab bar. It is not designed for this. The only grab bar suitable for a bathtub is a stainless-steel model secured to the studs. There are instruments that locate studs. If the company that you called cannot locate the studs, perhaps you should call another company.

Patching Plaster Walls

I recently installed a central-heating system in my home. The old heaters, measuring 20 x 60 inches, are mounted back-to-back in the wall between rooms. Removing these units will leave a huge pass-through between the rooms. I want to frame out the openings and repair the walls so the patch won't be noticeable. Should I use lath and plaster, or should I try to make a flush patch with easier-to-handle wallboard?

If you want a perfectly smooth wall you should cover the entire wall—from corner to corner—with wallboard. It's very difficult to achieve perfection with a patch. Depending on how light strikes the wall, you will see ripple shadows at the patched joints.

But, if you intend to hang pictures on this wall or cover it with a textured paint or wallpaper, patching would be adequate. Because of the size of the opening it would be easier to fill with wallboard than with plaster.

Roof-truss Uplift Cracks

There is a problem with cracks forming in my house. In cold weather, cracks appear at the corners where walls and ceilings meet. I had molding installed at the top of the walls two years ago, but this didn't help. What causes this, and how can it he repaired?

Roof-truss movement causes the seasonal opening and closing of the joint between the ceiling and the wall. This is commonly known as truss uplift. During the winter, the part of the truss immediately above the ceiling is exposed to higher temperatures and lower humidity than the parts of the truss immediately below the roof deck. These parts see lower temperatures and higher humidity. They pickup condensation that forms on the underside of the roof deck. As a result, the lower part of the truss tends to shrink slightly during the winter, while the parts immediately below the roof deck tend to swell. In turn this causes the upper parts of the

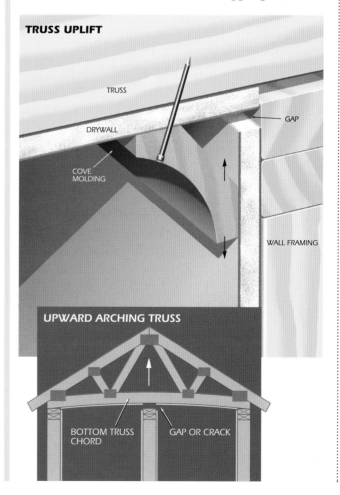

TRUSS UPLIFT

TRUSS

GAP

DRYWALL

COVE MOLDING

WALL FRAMING

UPWARD ARCHING TRUSS

BOTTOM TRUSS CHORD

GAP OR CRACK

truss to bend the bottom of the truss upward. This lifts the ceiling below, and cracks develop where the wall and ceiling meet.

This condition can be prevented with construction techniques, but the best way to deal with the problem now is with a cosmetic repair. You were correct to have moldings installed, but the moldings should have been nailed to the ceiling, not the wall (as is typically done). The moldings should be free to move up and down with the ceiling. As long as the molding is wide enough, it will cover the crack at the ceiling.

Creosote Removal

Will anything remove creosote from the brick wall behind my stove? Over the past 10 years the streak has lengthened, it and now reaches the floor.

If the creosote streak is thick, first tap it gently with a small hammer and pry it up with a putty knife. You can try to embrittle the streak, and make it easier to chip off, by applying a dry-ice pack to it.

Another option is to make a smooth paste from a solvent (such as benzene, naptha, or trichloroethylene) and an inert material such as talc. Remember when using these solvents to follow precautions printed on the container.

Smear the paste on the stained area with a trowel, allow it to dry and scrape it off. Repeat this as necessary. Afterward, clean the area with water.

Once you remove the streak, that area will be cleaner than the surrounding wall. To reduce the contrast between the streaked area and the wall, wash the entire wall with a cleaner and a stiff bristle brush. Then thoroughly rinse the area.

Painting Paneling

Is there a quick and easy way to rejuvenate the 20-year-old wood paneling in our family room?

The easiest thing to do is to paint the paneling. It's important to note that, in many cases, the paneling's color may have been achieved with a fairly soluble stain that can bleed through paint unless the surface is properly sealed. Therefore, the first coat should be an oil-based, stain-blocking primer. After the paneling has been primed, it can be top coated with latex paint.

Moldy Corners

Every winter I notice a dark mold that develops in the inside corners of our outside walls. I've tried scrubbing with detergent and repainting, but the condition returns in about two weeks. What's causing this and how can it be eliminated?

The dark mold is the result of mildew buildup. Apparently, those corners are colder than other parts of the wall and, consequently, condensation develops. The moisture accumulation encourages growth of mold spores.

Localized cold spots are usually the result of open joints or a thermal bridge. A thermal bridge is a building component that does not have good insulating value and directly connects an outside wall to an inside wall. Studs and headers can act as thermal bridges.

Most of a wall surface is exposed to warm-air circulation, which keeps the temperature above the point at which moisture will condense (dew point). The corners, however, have virtually no air circulation.

To control the mildew buildup, it's necessary to kill the mold spores. Detergent alone will not do the job. Scrub the area with a solution of detergent, water, and chlorine bleach.

Once the mildew has been removed and the area dried, repaint the wall with a quality mildew-resistant paint. Be aware that this type of paint generally contains fungicides that are poisonous. Do not use it on any surface that may be bitten by a child.

Another solution, although not very practical in this case, is to cover the inside surfaces of the exterior walls with ³/₄-inch-thick rigid-foam insulation board, which in turn is covered with drywall. This will raise the surface temperature of the wall above the dew point and thereby eliminate condensation and mildew.

Blistered Plaster

My problem is with the plaster in our house— it seems to be growing. It bubbles, almost like foam, and gets powdery. The problem is most noticeable under the windows.

Our house was plastered in extremely cold weather. Could the plaster have frozen while setting up, or is the problem due to moisture?

The problem is caused by a moisture condition. Constant wetting results in a breakdown of the crystalline structure of the plaster, causing it to swell. Your description suggests that the condition is active and must be corrected before repainting. In my experience, blistered and deteriorated plaster is always caused by water leakage.

Check for cracked or open joints around the windows where water can seep in. These should be caulked. Water dripping from the window pane due to condensation may add to the problem as could clogged drain holes at the bottom of storm windows. Once the moisture problem is corrected, scrape away all loose and deteriorated plaster. Then replaster the affected areas, prime and paint.

If the house was plastered when the interior temperature was below freezing, all the walls and ceilings would be soft and crumbly and you would have noticed the situation right away. When plaster freezes before it sets, there's no remedy but to remove it and replaster.

A Shrinking Problem

I just moved into a house that's one year old. There are open joints between some of the walls and floors which, I assume, are caused by shrinkage of the wood framing. What is the effect of the moisture content of wood on shrinkage and decay? Is the shrinkage of wood proportional to the change in its moisture content?

There is a direct relationship between the moisture content of wood and shrinkage. However, this exists only when the moisture content is below the fiber saturation point, which for most species of wood is about 30 percent.

Moisture content (MC) is expressed as a percentage that defines the weight of moisture-bearing wood

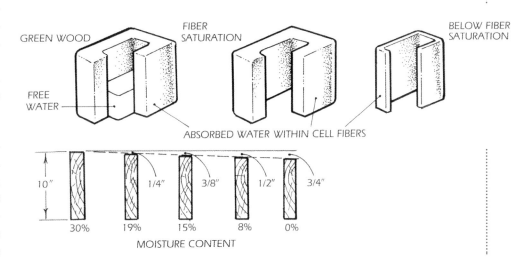

GREEN WOOD · FIBER SATURATION · BELOW FIBER SATURATION · FREE WATER · ABSORBED WATER WITHIN CELL FIBERS

10″ — 1/4″ · 3/8″ · 1/2″ · 3/4″

30% · 19% · 15% · 8% · 0%

MOISTURE CONTENT

relative to the same piece when oven-dry. Moisture exists in freshly cut (green) wood in two forms: as free water contained within the cell cavities, and as absorbed water contained within the cell fibers. The moisture content of green wood can be as high as 200 percent. Initially, as the wood dries, only the free water is given up.

The moisture content at which all the free water has been lost and all the absorbed water remains is called the fiber saturation point. Even though the moisture content has been reduced from 200 percent to 30 percent, no shrinkage occurs.

Shrinkage primarily occurs across the grain, and begins when the cell fibers give up the absorbed water. The total shrinkage that can occur takes place between fiber saturation point (about 30 percent MC) and a theoretical moisture content of 0 percent. For every 1-percent moisture loss below fiber saturation, there will be a $3^{1}/_{3}$ percent shrinkage (or 1/30 of the total shrinkage).

Depending on the temperature and humidity of the surrounding air, wood will give up or take on moisture until its moisture content is in equilibrium with the moisture in the air. Normally, wood never reaches a moisture content of 0 percent.

If wood is kept constantly in 70°F air with a 60 percent relative humidity, its moisture content will eventually reach about 11 percent. As far as decay is concerned, the decaying fungi (rot) will grow in wood only when the moisture content is above 20 percent. Moisture levels below this will not support decay, and the wood could last for hundreds of years.

WINDOWS

Sweating Windows

My windows sweat, and water runs down the window and onto the wall. I have storm windows and a gas-fired heating system. I sure would appreciate any advice you can give me.

There apparently is excessive moisture in the house's air, and there are several conditions that can cause this to occur. Two conditions that cause excessive moisture are also hazardous: a clogged chimney and a cracked heat exchanger in a furnace. Both water vapor and carbon monoxide are products of combustion in a gas-fired heating system. If the chimney flue is blocked or the heat exchanger is cracked, both carbon monoxide and water vapor will permeate the house. In a house that is tightly sealed, carbon monoxide, which is an odorless gas, could be deadly. You should have your heating system and its chimney checked by a competent technician. Another source of excess moisture is a malfunctioning humidifier on the furnace. Also, a dirt-floor crawlspace can allow moisture in the soil to migrate into the house. If this is the case, install a vapor barrier of polyethylene sheets over the dirt. Finally, bathrooms should be properly ventilated by an exhaust fan. The fan should discharge to the outside, not into the attic. In some cases, a blocked fan outlet can introduce excessive moisture into the house.

Damaged Insulated Window

Is there any way an insulated glass window can be repaired when the seal has broken and moisture has gotten in between the panes? I'd like to repair it rather than replace the entire window.

This is a frequently asked question. Unfortunately, the problem cannot be corrected by the homeowner. In many cases, not only is there moisture between the panes but there is also a discoloration that obstructs visibility through the pane. Even with this condition, the window will still serve an energy-saving function. It is more effective than a single-glazed window and is probably comparable to a storm window.

FIREPLACE

Smoking Fireplace

Three of the four fireplaces in our home work perfectly. But our largest fireplace has severe smoke backup when the wind is calm. When the wind is strong, the fireplace draws perfectly, and we don't have to crack open a window to maintain positive pressure inside the house. Incidentally, heating the chimney is not a solution, since the smoke backs up even during a hot fire when the wind is calm.

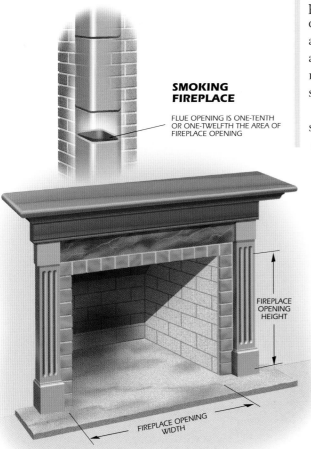

SMOKING FIREPLACE

FLUE OPENING IS ONE-TENTH OR ONE-TWELFTH THE AREA OF FIREPLACE OPENING

FIREPLACE OPENING HEIGHT

FIREPLACE OPENING WIDTH

There are three reasons why a fireplace will not function properly and cause smoke to back up into the room. The first is if you have negative pressure in your house. The second is if the chimney top is not high enough above your roof. It should be at least 3 feet above a flat roof and 2 feet above the ridge of a pitched roof. Also, the chimney top must be 2 feet above any part of the roof within a 10-foot radius of the chimney opening. Third, the fireplace opening must be properly sized to the area of the flue opening. Since the other three fireplaces work properly, it's unlikely that negative pressure is the problem, and the chimneys probably extend high enough above the roof, although this should be checked.

That leaves the fireplace opening to consider. When the area of the fireplace opening is too large for the flue opening, the fireplace will not draw properly. According to fireplace and chimney specialists, if the chimney is less than 35 feet tall, the area of the flue opening should be one-tenth the area of the fireplace opening. When the chimney is more than 35 feet tall, the area of the flue opening should be one-twelfth of the fireplace opening.

One way to achieve the correct ratio is to decrease the size of the fireplace opening by laying bricks on the floor and along the sides of the fire chamber.

Fireplace Backsmoking

I recently moved into a brand-new condominium that has a factory-built fireplace. Every time I use the fireplace, smoke fills the room rather than flowing up the chimney. Once the fire gets going, the chimney draws properly and pulls up the smoke. I don't understand why there's a problem. Can you help?

The problem is probably caused by a negative-pressure condition, also called depressurization, that exists in your house. This is a fairly common phenomenon in new construction, as opposed to in houses built 20 years ago. To conserve energy, new homes are better weatherstripped and caulked and are tighter than older homes.

All homes have a ventilation rate. This is the number of times the enclosed air volume changes in an hour. Air enters and leaves the house through various windows, doors and minute gaps. The typical house built 20 to 30 years ago had an average rate of 0.4 to 1.0 air changes per hour. Many new houses have an average rate of 0.1.

Depressurization results when more air in the house exhausts through fans and vents to the outside than flows in. The greater the exhaust and the tighter the building, the greater the depressurization. Also, because warm indoor air is lighter than colder outside air, it tends to leak into the attic through an access door or ceiling hatch, and then flows to the outside through vents.

When the damper is opened, there is generally an onrush of incoming air. In some cases, the rush of air is so intense it blows ashes into the room. To eliminate this, you must equalize the air pressure between indoors and out. Do this by cracking open a window or door, preferably on the side of the house on which the wind is blowing. If you open a door or window on the opposite side, more air will be drawn out of the house—creating an even greater negative pressure.

Overheated Gas Fireplace

My gas-fired fireplace shuts down, apparently from being overheated. The fire reignites when things cool off. I can't find a damper to let some of the excess heat escape.

It sounds like you have a vent-free fireplace. Gas fireplaces have either a direct vent through the wall of the house, a natural vent up a chimney, or they have no

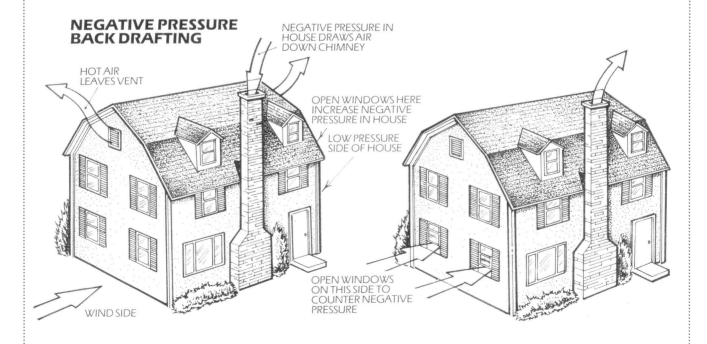

NEGATIVE PRESSURE BACK DRAFTING

NEGATIVE PRESSURE IN HOUSE DRAWS AIR DOWN CHIMNEY

HOT AIR LEAVES VENT

OPEN WINDOWS HERE INCREASE NEGATIVE PRESSURE IN HOUSE

LOW PRESSURE SIDE OF HOUSE

OPEN WINDOWS ON THIS SIDE TO COUNTER NEGATIVE PRESSURE

WIND SIDE

vent, in which case they are called vent free. Vent-free fireplaces have no need for a damper. These appliances have a thermal sensor located near the top of the fireplace opening. This is a safety device that in a properly operating fireplace would never be triggered. The sensor cannot be reset. It comes with the unit and is set at the factory.

Some vent-free fireplaces also have an oxygen depletion sensor. If the fire is taking too much oxygen out of the room's air and it's not being replaced, the sensor will prevent combustion. Either one of the above sensors could be defective and cause the problem that you describe. You should contact your gas supplier or a hearth-products shop to find a technician to repair the problem.

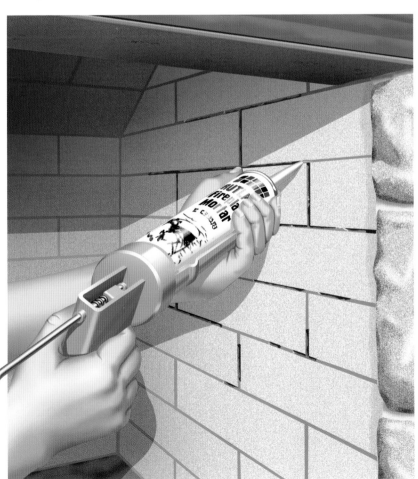

Fireplace Mortar Joints

Our home was built in 1957. It has a conventional wood-burning fireplace in the living room. After all these years, some of the bricks are missing mortar in many places at the back of the fire area. Can you recommend what type of mortar would work best in this high-heat application?

You have two options. One is to buy a ready-made mortar in a cartridge. The other is to mix your own to industry-accepted standards. If you don't want to mix your own mortar, you can apply a ready-mixed product, available at fireplace stores, that comes in a cartridge and is applied with a caulking gun. This mortar has high temperature resistance, and it requires heat to be cured to its final hardness.

If you decide to mix your own mortar, you should be aware that there are many different types, and each is designed for a specific application. According to the Brick Industry Association (BIA), either Type N or Type 0 mortar would be a good choice. There are two types of Type N mortar. One consists of one part portland cement, one part hydrated lime, and six parts sand. The other consists of one part Type N masonry cement and three parts sand. Type 0 consists of one part portland cement, two parts hydrated lime, and nine parts sand. If you use premixed mortar in a bag, check with the manufacturer to see what industry specifications it meets.

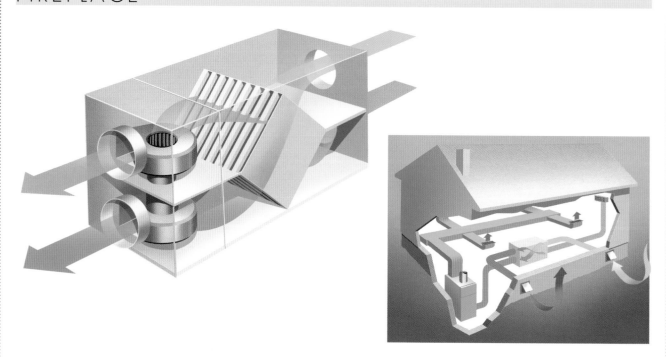

Equalizing Negative Pressure

I have a smoking fireplace, and I have had little success getting help. I've checked the chimney for obstruction and proper height, and I have also checked the size of the fireplace opening. All of these are okay. This leaves negative pressure as the last possible cause. We do not want to have a window or door open. We already have a problem with soot getting on our white drapes, white carpet, and white furniture.

The easiest way to equalize negative pressure in a house is to open a window or two. However, since that's the one thing you don't want to do, I would recommend that you install a heat-recovery ventilation system. The heart of the system is an air-to-air heat exchanger. These units are primarily used for exhausting indoor air pollutants and maintaining good indoor air quality. Note in the drawing that the heat-recovery core uses the heat in the stale indoor air to heat the fresh cold air that enters from outside.

There are no moving parts and no refrigerant in the core. It is made from polypropylene plastic. The air simply is drawn through the core by the fans, and the core is warmed by the indoor air. The two air streams do not come in contact with one another. A switch on the side of the ventilator allows the homeowner to control it, and its fan speed is adjustable.

Research has shown that the dark staining that occurs on the edges of carpeting, draperies, and bed ruffles is more often caused by soot originating inside the house rather than dust or dirt from outside. The soot may come from a fireplace, but the condition also occurs in homes without fireplaces. It may come from furnaces, standing pilot lights, candles, and cigarette smoke. With an air-to-air heat exchanger, some soot will be exhausted.

A heat-recovery ventilator can use the ducts of a warm-air heating system, or a system can be installed in houses heated by hot-water radiators, radiant panels, or electric baseboards. The system also can be used during the summer months.

Stained Bricks

I made the mistake of putting a potted plant (with a dish under the pot) on the red bricks in front of the fireplace. There is now a round water stain left by the dish. Is there a way to remove the stain? Thank you in advance for your help.

Water stains on brick are difficult to remove. But here's something worth trying: Make a solution of laundry detergent and water, then add some bathroom talc to form a paste. Smear the paste on the stained area with a trowel. After the paste has dried, scrape it off. Repeat the process if necessary.

Wood Stove Clearances

To prevent a fire it is important that proper clearances be maintained between wood stoves and combustibles such as walls and furniture. Wood undergoes a physical change when continually exposed to elevated temperatures, and this reduces its ignition temperature. Wood normally begins to burn at 400° to 600° F. However, when it is continually exposed to temperatures between 150° and 250° F, its ignition temperature can be lowered to 200° F.

It can take years for wood's ignition temperature to be lowered, so homeowners can acquire a false sense of security. But when the wood's temperature coincides with the lowered ignition temperature, it will ignite and burn spontaneously.

With that in mind, there are basically two sets of standards for clearances around wood stoves—clearances for stoves that have been tested and listed to UL Standard No. 1482, and clearances for stoves that have not been listed to this standard.

A new stove will have a label attached to it telling you if it has been tested and listed to this standard. The label will also tell you what clearances to use around the appliance (generally these are in the 12- to 36-inch range).

For stoves that have not been tested and listed to the UL standard, the minimum clearances are often defined by local building codes, so you should check with your building department before installing a stove. Also, you may need a building permit for the installation, and some municipalities require that a stove meet current EPA emissions guidelines.

BATHROOM

Sewer-gas Odor

Eight years ago, we remodeled our bathroom and installed a vanity sink with a faux marble top and no overflow hole. After a while, a terrible odor began building up and getting worse. We replaced the sink top two years ago (this one has the overflow hole) and again the odor is terrible. We've kept the drain clear with drain cleaner, but that doesn't seem to affect the odor. Do you have any ideas about its cause?

The problem is not related to the presence or absence of an overflow hole in the sink basin, or the need for a drain cleaning product. Although you didn't mention the type of trap in the drain line, I suspect it is an S-shaped trap rather than a P-shaped trap. The former is commonly called an S trap, the latter a P trap. An S trap could cause the problem you are experiencing. Most municipalities no longer permit the installation of S traps in new construction. However, during my home inspections over the years, I have seen many of them in older houses as well as a few in new homes.

The purpose of a trap under a sink is to prevent the escape of sewer gas into the room. It does this by trapping water in the U-shaped portion of the pipe, forming a water seal. The water blocks the movement of sewer gas and prevents it from flowing up through the trap, the sink drain, and into the room.

The problem with an S trap is that the water seal may be lost due to siphonage. In the case of an S trap, the pressure of air on the water in the fixture is greater than the pressure of air in the drainpipe on the down side of the trap. The action of the water discharging into the drainpipe removes the air from that pipe and thereby causes a negative pressure in the drain line, resulting in siphonage of the water from the trap. The water seal is lost, and sewer gas can flow into the room. If this should occur, and you smell sewer odors, an easy fix is to open a faucet and let the water run for about 5 seconds, which is enough to fill the trap.

Siphonage is unlikely when a P trap is used. The horizontal portion of the trap is connected to a plumbing vent pipe which extends through the roof, so there is always atmospheric pressure on the down side of the trap.

P TRAP STOPS SEWER GASES

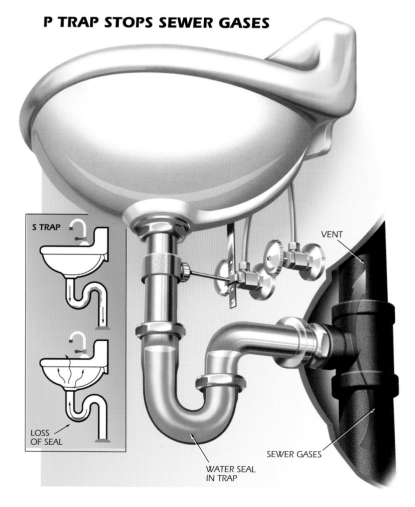

S TRAP

LOSS OF SEAL

VENT

SEWER GASES

WATER SEAL IN TRAP

Antiscald Measures

We have a problem with our hot-water system. If someone is taking a shower and another person runs water or turns on a water-using appliance, the water in the shower becomes so hot it can scald the person who is bathing. What can we do to prevent this?

There are two things you can do to prevent this problem. First, lower the temperature of the hot water. If the hot water is generated in a tank-type water heater, lower the thermostat setting until the water discharging from the faucet is no hotter than 120°F.

If the hot water is generated by a tankless coil in a boiler, install a coldwater mixing valve between the coil inlet and the hot-water outlet. The valve should be adjusted so that the temperature of the water delivered to the fixtures does not exceed 120°F.

Another option is to replace the existing hot/cold shower valve with an antiscald valve, available at plumbing-supply stores. These valves work on the principle of pressure balancing. When the cold-water pressure is reduced, for whatever reason, the hot-water flow is also reduced, thereby maintaining a relatively constant temperature. However, to install this valve you will need to open either the front or back of the shower wall. In many cases, this involves removing tile and the substrate below it, then replacing both. It amounts to a small, but expensive, remodeling job.

Phantom Flush

Every day, our toilet makes sounds as if it were being flushed, when no one is near it. What could be causing this to happen? Is there something we can do to fix it, or do we need to call a plumber?

This is a common problem, although it can be quite unnerving if you hear it for the first time in the middle of the night. You might think you have an uninvited guest in the house. The flushing sound is not an actual flush. It's water refilling the toilet tank. Possibly, there is a flush valve in your toilet allowing a small amount of

LEAKING FLAPPER VALVE

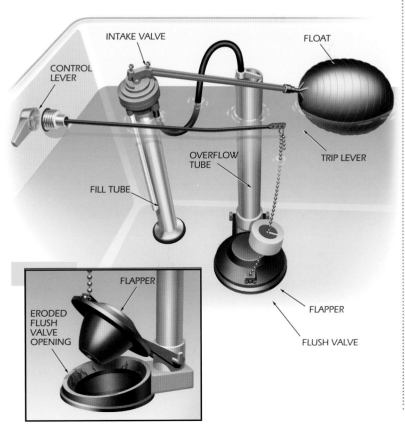

CONTROL LEVER

INTAKE VALVE

FLOAT

FILL TUBE

OVERFLOW TUBE

TRIP LEVER

FLAPPER

ERODED FLUSH VALVE OPENING

FLAPPER

FLUSH VALVE

water to drain from the tank. You don't hear the water flowing through the valve into the bowl, but after several hours, the tank's water level drops to a point where the water fill valve is actuated, and the tank is refilled with water.

The leak could be the result of improper alignment of the flapper on the flush valve seat, a slight crack or deterioration of the flapper, or possibly a dirt buildup on the valve seat. If you are handy, you won't need a plumber to correct the problem. This is a typical do-it-yourself job. Shut off the water valve to the toilet, flush the toilet, then empty the water remaining in the tank. Replace the flapper, and clean the valve seat. The job should take less than an hour.

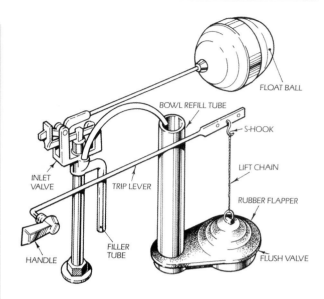

Slow-flushing Toilet

My toilet suddenly started acting up: I either have to hold down the handle for three seconds before it flushes or snap it really quick to flush. I prefer to do it the way everyone else does—just flush. What can I do?

There is probably too much slack in the lift chain. When you flip the handle, the rubber flapper lifts only slightly off the flush valve seat. The pressure of the water on the flapper causes it to reseat on the flush valve, stopping water flow.

By holding the handle for three seconds you are holding the flapper off the valve seat so that the water will flow into the bowl. The water rushing past the flapper holds it up after you release the handle. It also flushes when you snap the handle quickly because the jerky action forces the flapper up to a point where water flowing into the flush valve holds the flapper open.

There are several causes for a slack lift chain. The chain may be slightly rusted, the hook connecting the chain to the trip lever may have stretched or the flapper may be deteriorating.

Regardless of the cause, the correction is simple:

eliminate the excessive slack in the chain. Do this by moving the chain hook to a different hole on the trip lever or replacing the lift chain. Replace the flapper if it shows excessive wear.

Rocking Toilet

We have a toilet that is loose on its base. I have tried tightening its hold-down nuts, but that doesn't seem to help. It's embarrassing when we have company and our guests have to use this rocking toilet. Please help.

Since you have already tried tightening the nuts, and that didn't stop the rocking, you'll have to check the area under the bowl. To do this, shut off and disconnect the water supply, flush the tank and sponge out residual water. Remove the closet nuts and lift the bowl off the closet bolts.

Three things can be causing the problem. First, the closet flange may be loose or rusted out. Second, the subfloor around the flange may be rotted out, in which case that section needs to be replaced. Finally, the wax ring and its sleeve may be flattened to the extent that it no longer bears properly against the bowl's base, and therefore a new wax ring and sleeve must be installed.

CLOSET NUT

CLOSET BOLT

CLOSET FLANGE

WAX RING

CLOSET NUT
AND BOLT
WASHER

WAX
RING

BOWL
FOOT

FLOOR

CLOSET
FLANGE

SLEEVE

SOIL PIPE

Deliming Toilet Jet Holes

The rim holes on my toilet have a scale buildup that is causing the bowl to flush more slowly than it should. I have heard of using acid to remove this scale, but is it safe?

I don't recommend that homeowners use a high-strength acid to remove scale buildup.

Instead of high-strength acid, try swabbing under the rim with full-strength white vinegar, which is generally a 6 percent solution of acetic acid and water. If the scale buildup is not too thick, several applications should do the job. The scale accumulates on the outside of the jet holes because a film of water adheres to the surface of the porcelain, and when the water evaporates, the minerals that were in it are left behind. Don't put the vinegar in the toilet tank—the scale is on the outside of the rim holes and flushing it through will not be as effective as the method described above.

If vinegar fails, use a scale-removing household cleaner, but it's important to follow the product's directions in order to work safely with it. These cleaners are sold in supermarkets and home centers.

Sweating Toilet

Why does our toilet tank sweat? It gets so bad, a puddle forms on the floor. Leaving the lid up or down doesn't seem to matter. Do you have any suggestions?

A toilet sweats because its surface is cool and moisture in the air condenses on it. The wall of the toilet is cooled by the water standing in the bowl and the tank. There are two ways to prevent this problem. One way is to install insulation available in kit form at hardware stores and home centers. The insulation is rigid foam, and it's installed with a little adhesive along the walls on the inside of the tank.

By preventing the cool water from touching the wall of the tank, the insulation keeps the tank's surface temperature warm enough to prevent moisture in the air from condensing on the tank. These kits do not solve the problem of condensation that forms on the bowl, however. The other method is to have a plumber install a valve and pipe to introduce a small amount of hot water into the cold-water feed line to the tank.

Improper Flush

My 13-year-old house has three toilets, two of which flush properly. The third does not seem powerful enough to flush solids. I've changed some parts and made some adjustments, none of which seem to help.

My initial thought was that the toilet was an early 1.6-gallon model. Federal water-conservation laws required the installation of these toilets beginning in 1994.

However, they soon become synonymous with poor flush quality. Newer, redesigned models deliver a more reliable flush.

But I'll assume that the toilet is an older 3.5- or 5-gallon type. First, check the tank's water level. It should admit water to the fill line marked inside the tank. If it doesn't, adjust the flush valve. Also, check that the flapper opens properly. Next, use a small mirror to check that the rim holes are not blocked by mineral encrustation. If this is the case, they can be cleaned out with a small nail. You may also have to use an acid based porcelain cleaner.

Finally, check that the vent stack that serves the toilet is clear. Water and waste displace air while moving through a building's drain system. Vents provide a means of letting in makeup air and they also provide an outlet for harmful waste gas. Air enters a house's drain system by way of a vent stack, a pipe that exits through the roof. In the case of a toilet, a lack of air can cause improper flushing action and a variety of other problems, so the vent stack that serves the toilet must be checked for obstructions. If the vent stack is blocked by a bird nest, leaves, or even a dead animal, then adequate air will not be allowed to enter the drain system. The only way to check the vent stack is to go up on the roof. Whether you decide to do this depends on how comfortable you are working on the roof and if you think you can do this safely.

Low-flow Bowl Problems

I recently modernized one of my bathrooms and installed a 1.6-gallon toilet. But the toilet often doesn't adequately dispose of solid waste in one flush, necessitating a second and sometimes a third flush. What can I do to alleviate this?

The first thing you should check is that the flush valve is properly adjusted. Although there may be an adequate amount of water in the tank, if the flush valve is improperly set, the correct amount of water will not flow into the bowl. The bowl's installation instructions give information about the proper setting. If you try altering the flush-valve setting and this does not help, contact the company's customer service department.

Falling Water Causes Back Pressure

Occasionally the water in my toilet bowl gurgles, and other times it is forced out onto the floor. This doesn't occur when the toilet is flushed but happens at random moments. The toilet works as it should in all other respects. What is the source of the back pressure that is causing this? I live in a condominium apartment on the ground floor of a building several stories high.

A back-pressure condition is not uncommon in mid- and high-rise buildings, and it occurs in tall houses that have a bathroom on the top floor. The fixtures in which this occurs are those at the base of a soil stack or where a soil pipe changes its direction abruptly.

This is how the back pressure occurs: A large amount of water, usually from a flushing toilet, will discharge into the soil stack. The water falls through the soil stack (sometimes described as a slug of water), and the air in front of it becomes pressurized because it is unable to slip past it and out the roof vent. Since the sewer pipe slopes, the airspace tapers to a point, which traps the air in the sewer pipe. The greater the distance that the water falls, the greater the pressure. This is why the problem is found in mid- and high-rise buildings and tall older homes that have had a bathroom added on the top floor or in the attic.

Incidentally, this does not occur when a sink drains into the soil stack. This smaller volume of water tends to cling to the wall of the soil stack and spiral its way down

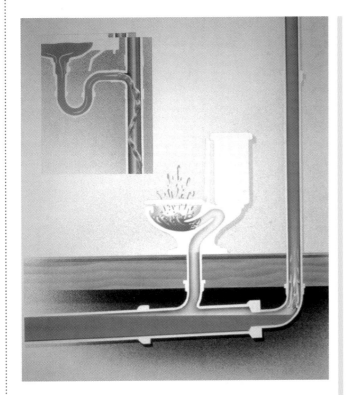

cleaners that are formulated specifically for removing mineral stains. These cleaners are sold at supermarkets and hardware stores. One application is usually enough to remove normal amounts of mineral deposits. Heavy deposits may require more than one application.

Painting Tile

Is it possible to paint ceramic tile that is used in a bathtub/shower area? We recently installed new flooring, sink top, and vanity in our bathroom, and now the orchid tile (circa 1960s) looks terrible.

I'm sorry to tell you that there is no way to permanently change the color of tiles by painting them, especially tiles located in a bath area. An epoxy paint will form a reasonable bond with the glazed surface. However, after repeated cleaning, which is usually necessary to remove mildew and soap scum, the surface will begin to wear or peel.

Peeling Paint in the Shower

I have had a continual problem with the paint in the shower peeling and flaking off. I've been using a concrete water-repellent paint which lasts for a period of time. What type of paint would you recommend to use in a shower with a poured-concrete base and conrete walls? Should I place tile over the walls? The house was built with this type of shower in 1946.

I believe you would be ahead of the game if you tiled the walls and floor of your shower rather than painted those surfaces. Tiles may cost more initially, but in the long run they will be less of a problem. They also

the stack as it falls. It does not pressurize a column of air as does a large volume released from a single flush.

Your toilet is probably close to the base of the soil stack, and if the pipe serves a number of bathrooms on the floors above, the pressure developed can cause the air to bubble up through the bowl water, causing the gurgling sound. If the pressure is great enough, it can blow the water out of the bowl.

Porcelain Stains

The high mineral content in the water where I live stains porcelain surfaces, such as the entrance to a toilet's trap. These dark stains are very difficult to remove, even with bleach. What else can be used to remove them?

This is a common problem in areas where there is a high mineral content in the water. Fortunately, the dark crust formed by these minerals can be removed using provide you with a wide selection of decorative patterns, which are not available in a painted surface.

The peeling and flaking of the paint in this type of application is not unusual. Quite often the surface to be painted is not properly prepared. The surface must be clean and dry. All soap residue, dry salts, and oily films on the surface must be removed. Depending on how much of a problem is presented, the cleaning may have to be done by washing the wall with a dilute solution of muriatic acid, possibly by mechanical scrubbing by means of a belt sander or wire brush.

The unpainted concrete surface may feel dry to the touch but may still contain a lot of moisture as a result of being wet from the shower.

An easy test to determine whether the wall is dry is to tape an 8-inch-square section of aluminum foil to the wall, sealing the perimeter with masking tape. Remove the foil 24 hours later and see if the side facing the wall is damp. If it is, that means the wall surface is damp and should not be painted.

Even when precautions are taken, periodic repainting of the shower walls and floor should be anticipated. On the other hand, with a tile surface the periodic maintenance is simply regrouting cracked tile joints and occasionally resetting a loose tile. Tiling is your best bet.

Sink Overflow Odor

I have a 2½-year-old house. About a year ago, the sink in the master bathroom started to develop a moldy odor. I tracked it to the overflow hole at the top of the sink basin. I tried to get a brush down the hole, but I cannot get one to fit. I have tried a foaming-type bathroom cleaner, but that does not affect the smell. The only thing I can do is pour some bleach down the hole, but that only stops the smell for about three days. What can I do to get rid of the smell permanently?

The bleach that you are pouring into the overflow hole is just flowing down the drain. It removes the surface mold, but not the mold buildup. In order to kill all of the mold spores, you have to prevent the bleach from draining out of the overflow channel.

To do this, first remove the sink's pop-up stopper. Either lift the pop-up out directly or turn it counterclockwise and lift it out. If you have an older sink, you probably will have to remove the pivot rod and its assembly in order to remove the pop-up.

Once the pop-up is removed, stuff a rag down the sink drain hole. If you had to remove the pivot rod assembly, you should also place a pan under the sink to catch any runoff from the cleaning process.

Next, open a window to provide ventilation. Then, pour a mixture of 50 percent bleach and 50 percent water into the overflow opening. Use a funnel to do this, and stop when the mixture overflows the opening. Some of the mixture will leak through the rag and into the drain, so periodically pour more of the mixture into the overflow opening. Let the mixture stand for 30 minutes, then flush the overflow channel with clear water. Be sure to clean the pop-up stopper with the mixture before reinstalling it.

Hairline Cracks

Last year we had a beautiful cultured-marble countertop and sink installed in our new bathroom. Just lately, we noticed tiny cracks that have spiraled out from the sink strainer. The cracks are thin, but getting longer. They do not go through, and are limited to, the surface. Have you ever heard of this? What can we do?

The surface cracks near the drain can be caused by very hot water. However, more often than not it is the result of overtightening the drain fitting with a wrench rather than making it hand tight. Unfortunately, the condition cannot be repaired.

Faucet Squeal

The hot and cold faucets in the upstairs bathroom tend to squeal when they are halfway open. Turning them to the full-open position stops the squeal. What's the fix?

Although you didn't mention it, I assume you are referring to a washer-type faucet. A high-pitched sound from one of these faucets is usually caused by water flowing at a high velocity through a narrow constriction. Fully opening the faucet eliminates the constriction and the squeal.

It is possible that this is caused by a cut in the faucet's washer, which could have been caused by a burr on the faucet seat. A loose washer can also cause the noise, although this usually results in a chattering sound rather than a squeal. Try replacing the washer. If the condition persists, you should reface or change the seat. Visit an old-fashioned hardware store or plumbing-supply house to get the tools you need.

Gold-plated Fixtures

We recently moved to a home that has gold-plated bathroom fixtures. Evidently they had been cleaned with an abrasive solution because all the finish is gone. Is there anything we can do to restore their gold luster?

The finish on gold-plated faucets resists soap and hard water. However, it won't withstand an abrasive cleaner. Unfortunately, there's nothing you can do to restore the finish other than replating the parts, and I doubt that the cost would be justified. Unless the finish on the entire faucet has been abraded, you might consider replacing just the affected parts rather than the entire faucet. Contact your local plumbing-supply store for assistance.

Faulty Shower Diverter

I have a problem with the shower diverter on my bathtub faucet. When the knob is pulled up to divert water to the shower, a lot of water continues to flow into the tub. Is there any way to either replace or repair this unit without going into the wall?

Yes, there is. The diverter valve mechanism that you refer to is inside the tub faucet spout, not inside the wall. The spout is screwed onto the water pipe and can be unscrewed easily. Once the spout is removed you can see the diverter mechanism. The diverter is a small gate valve attached to the base of the plunger shaft. It is held up (closed position) by water pressure. When the water is turned off, the gate valve drops and opens the tub spout.

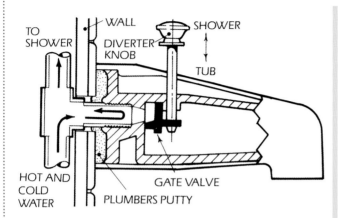

If the diverter can't be repaired, it's cheap enough to replace the entire spout, including the diverter mechanism. These spouts are available at plumbing supply stores. There are different types and sizes of diverter spouts available, so take your old spout along to the plumbing supply store to make sure you get the right one. Check especially that the set-back distance of the threads within the spout matches the length of your protruding water pipe, so you get a tight fit between the spout shoulder and the wall.

When replacing the spout, use pipe-joint compound on the pipe threads. Completely fill the hollowed-out back end of the spout with plumbers' putty to prevent water from penetrating the wall.

Tub/Shower Diverter

The water pressure in our tub is good. However, when we turn on the shower, some water continues to come out the tub faucet, and the water pressure from the showerhead is reduced. How can we correct this problem?

The problem is with the tub/shower diverter valve. When the valve is functioning properly, all the water should be channeled to the tub or the shower.

Two common types of diverters are shown in the diagram. In the tub spout diverter, a small gate valve is attached to the base of a plunger shaft located in the spout. It is held in the up position for the shower by water pressure. When the water is turned off, the gate drops and opens the tub spout. To remove it for repair, unscrew the tub spout from the threaded water pipe.

The other diverter works by rotating the valve stem so the parts in the valve body open to either the tub spout or the shower pipe. You can remove the mechanism by unscrewing the stem nut and withdrawing the diverter assembly.

Tub diverter spouts and diverter assemblies can be purchased at plumbing-supply stores. Since there are different types and sizes of tub/shower diverters available, take the old one when you go to the store to make sure you get an equivalent replacement.

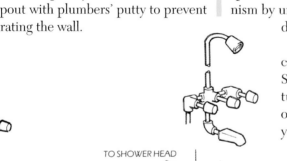

Aluminum Corrosion

How can corrosion (white rust) be removed from aluminum? Our shower doors were shiny aluminum, but over the years they have stained. White streaks and spots have formed on them. I have tried aluminum cleaner (phosphoric acid), acetone, paint remover, alcohol, and a variety of other agents with no success.

The white rust is aluminum oxide. If the aluminum frames on your shower doors were shiny, they probably had an anodized finish rather than a mill finish. The anodized finish is a chemically grown oxide film that makes aluminum smooth, very shiny, and protects it from corroding. An anodized coating can last for many years. A mill finish, on the other hand, is generally not used in places where appearance is important because it tends to turn a dirty gray and requires frequent cleaning to maintain its bright appearance.

Normally, all you need to clean anodized aluminum is a damp rag. If this doesn't do the job, then the finish has been damaged, and the shower-door frame is no longer protected from corrosion. The anodized coating may have been damaged when soap scum was cleaned off the door. Homeowners often use a steel-wool pad or strong acidic cleaner to do this. This will damage an anodized surface and lead to increased corrosion.

Although the finish can't be restored, you can clean off the white rust using a plastic 3M Scotch-Brite abrasive pad, and coat the frame with a waterproof clear varnish. This will give it some protection, but it's not nearly as effective as the anodized finish.

Green Water-Staining

There is a green stain on the wash basin and also on the tub in my summer home. The water is clear, but it leaves this stain. Would you know what causes this? Are the pipes too old?

The problem is not caused by the pipes, but by the water. Probably, the water is soft and has a high carbon-dioxide content, which makes it slightly acidic. This water leaches small amounts of copper from the pipes.

The stains are left from water that drips from a faucet. Each droplet evaporates, but it leaves a copper residue, and when the residue reacts with the air, it turns green.

Remove the stains with a weak acid on a sponge. Try using lemon juice or vinegar. The stains can also be removed by gentle cleaning with a mild scouring cleanser.

Leaking Fan Duct

I built my own home two years ago and installed an exhaust fan in my bathroom. My problem occurs when warm moist air meets the cold dry air via the flexible duct. The water pours out of the fan duct, which is vented in the ridge vent.

As you know, moist air condenses when its temperature drops below the dew point. Usually the exhaust fan can move the warm, moist air fast enough so that its temperature doesn't reach this point.

In your case there is enough heat loss along the duct to cause the moist air to condense. Your problem can be solved by insulating the duct. This reduces the heat loss and keeps the temperature of the moist air above the dew point. If you can, try reducing the flexible duct length. This will result in a higher temperature of the discharging air, which may be enough to eliminate the condensation. If not, you'll have to insulate the duct.

Fan Overload

I have a problem with my bathroom exhaust fan that I hope you can help me solve. I installed the fan myself in a 5-foot × 9-foot bathroom. Although the fan is designed to handle a space bigger than mine, it takes up to 15 minutes to clear out the air. I have a flexible tube to vent the unit, and the discharge end checks out okay. The bathroom door fits snugly in its frame, and against the living room carpet.

The answer to your problem is in the last sentence of your letter. An exhaust fan works by creating and holding a slight negative pressure in a confined space. However, to operate properly, the fan requires air to enter the room. Since your door fits tightly all around, there's no way (assuming there's no window) for air to enter the room. The easiest solution is to leave the door slightly open.

If you want to keep the door closed, then you can cut about $1/2$ inch off the bottom of the door. In a typical 30-inch-wide door this is equivalent to a 3- × 5-inch opening. If you do have a window, opening it slightly will create the same effect.

Of course, it's always possible that you installed the exhaust fan backward and it's blowing air into the room. Check this by holding a tissue up to the fan. It should be held in place on the grille.

Exhaust Problems

The fans in my bathrooms discharge into my attic, which is insulated and vented at each end with louvered gable vents. Will this system cause the insulation to become damp? Should I do anything about it?

Discharging moisture-laden air into your attic can cause condensation problems, especially if the attic isn't adequately ventilated. In my experience, most vented attics remain inadequately ventilated because insect screens and louvers reduce the effective vent openings. If there's a vapor barrier on the attic floor, then the unobstructed vent opening should be 1/300 of the floor area. Double the required vent size if there is no vapor barrier.

The most effective way to solve your problem is to discharge the moisture-laden air outside the house. A flexible duct connected to the fan and run under the insulation to an exterior wall where a hole can be cut will do the job. Install a weatherproof vent hood over the exterior wall opening. Attaching the flexible duct to an existing attic vent opening is not recommended because the duct will partially block the opening.

Another way to terminate the duct is to run it directly up through a hole in the roof. In this case, the duct should be as short as possible to prevent condensation inside the duct during the winter months. A suitable vent hood must be installed and the joints between the hood and roof must be sealed to prevent leakage.

KITCHEN

Gurgling Sink

My husband replaced a 15-year-old sump pump, and we now have a gur-gling noise in our kitchen sink every time the sump pump goes off. The old sump pump was $1/3$ hp, and the new one is $1/2$ hp. The sump pit is located in our basement just below the kitchen sink. Water from the sump pit and the kitchen sink discharge into the sewer line. We never had this gurgling noise before, and we want to stop it.

The gurgling noise is probably the result of your sink drain not being properly vented. Venting is very important to your house's plumbing. It serves two functions. Venting provides atmospheric pressure within the drainpipes, and this prevents water from being sucked (siphoned) out of the sink traps—a condition that would allow sewer gas into the house. Venting also lets sewer gas escape into the atmosphere.

A sink trap must be connected within a short distance to a vent stack. This is usually done with a P-trap. S-traps, which were used in the early days of plumbing, are not permitted in most municipalities because they are not vented. Consequently, the water seal in the trap can be siphoned off.

When your new, more powerful sump pump discharges into the house's sewer pipe, the high volume of water rushing past the sink drain connection draws air out of the drain pipe and creates a negative pressure. This negative pressure is strong enough to pull the water out of the trap, creating the gurgling noise.

You can correct this problem by installing an automatic vent valve on the downstream side of the sink trap. This value equalizes the pressure by letting fresh air in, and closes down when the drain flow stops. Before installing one, check with your local building department for approval.

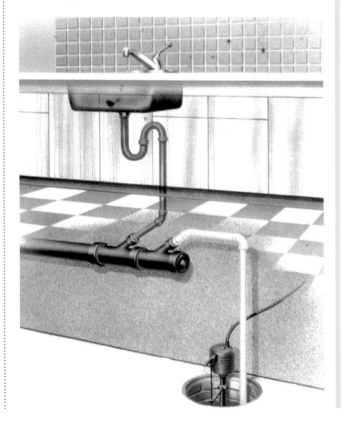

You also mention that water from the sump pit is pumped into the sewer line. Although this connection is not permitted in most municipalities, I've seen it hundreds of times while doing home inspections for prospective buyers. Municipalities often don't permit this because it introduces rainwater into the sewer line where it flows along with raw sewage to the treatment plant for processing.

Treatment plants are designed to process a certain quantity of sewage. By introducing rainwater into the sewage flow, homeowners unknowingly reduce the sewage treatment capacity, of' the plant. In order to treat all the sewage that is produced, additional treatment plants are built, and this causes taxes to increase.

Some municipalities have combination storm and sanitary sewers and allow sump pumps to discharge into the sewer line. However, unless this is the case where you live, redirect the pump discharge line so it spills onto your lawn. Just make sure it's far enough away from your house's foundation so that the water doesn't seep back into the basement.

Venting Island Sink

I'm planning some improvements to my kitchen that include locating the sink in a central island cabinet. I'm unsure of how to vent the drain. Do you have any ideas?

I agree that proper venting of the sink drain is important. Otherwise it's possible for the water seal, normally maintained in the drain trap, to be siphoned out by water flowing rapidly down the drain. If this happens, sewer gases will escape into the room.

There are many ways to adequately vent an island sink. However, before installing any system, check with the building department in your area to see if your design complies with the local building codes.

The Residential Plumbing Inspector's Manual, published by the American Society of Sanitary Engineering,

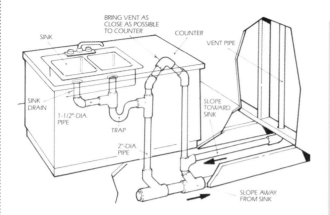

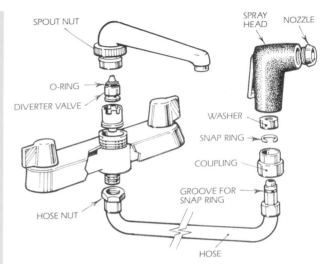

offers an arrangement that may be suitable for your needs (see drawing). Install the vent piping under the counter as high as possible. Then, return it downward and connect it to the horizontal sink drain immediately downstream from the vertical drain section. The returned vent is connected to the horizontal drain and to the vent pipe immediately below the floor with Y-branch fittings. The vent pipe should be carried up through the nearest partition or be connected to other vents.

Drainage fittings should be used on all parts of the vent below the floor and a minimum slope of ¼ inch per foot back to the drain should be maintained. The return bend under the counter can be a one-piece fitting, or assembled with 45 degree and 90 degree fittings. The island sink drain, upstream of the returned vent, should serve no other fixtures.

Sink Sprayer Problems

We have a kitchen faucet with a spray attachment. When we try to use the sprayer, only a light stream comes out with no force and most of the water comes out of the faucet. Can you tell us how to fix the sprayer?

First, look beneath the sink to see if the hose is kinked. If it is, and the kink can't be corrected, then replace the hose. Make sure you get either an identical

replacement or one with an adapter to fit your faucet.

If the hose is okay, then remove the sprayer head nozzle and clean it if necessary. The nozzle is usually screwed into the spray head. With the nozzle off, flush the spray head by turning on the water and depressing the spray handle.

If you still have a problem, then it's time to check the diverter valve. First, remove the faucet spout. The valve can then be lifted or screwed out and cleaned. The diverter area in the faucet body should be flushed at this time by slowly opening one of the faucet handles. If the problem is not corrected after you reinstall the diverter valve, the valve is defective and should be replaced.

Varying Hot-Water Pressure

The hot water in my kitchen faucet and a basement faucet are flowing slowly, but the flow in a second-floor bathroom is strong. What could be the problem?

There is evidently a constriction in the water supply line leading to those sinks. Check the hot-water valves under the sinks—they may be closed. Try to trace the hot-water supply to the sinks with the low flow. They may share a common section of pipe that is not in the loop with the other sinks. If they do, look for a partly closed valve, sharp bend or kink in the pipe. This will reduce the water flow. If the section of pipe is iron, it may appear to be in good condition on the outside, yet have a buildup of mineral deposits and rust on the inside, which reduces flow. if none of these are causing the problem, then I would suspect that the hot-water faucets are faulty.

Shiny Finish on Laminate

Due to misuse and haphazardness, we have removed the shiny finish from parts of our Formica kitchen counter. The counter is still in good condition, but we'd like to know if we can restore the original finish.

Unfortunately, there is no way to restore the shiny finish. The laminate is formed under heat and high pressure—conditions that are impossible to duplicate in the home. Once abraded, the finish can't be restored.

If you are unhappy with the finish of the countertop, you can resurface it with new laminate—provided the existing laminate is sound, well bonded, and does not have deep texture.

You can call the Formica Corporation for an instructional sheet on resurfacing laminated assemblies. If you are interested in cleaning and maintaining Formica plastic laminates, ask for the free brochure, "Caring for Formica Brand Laminate Surfacing Material."

Countertop Repair

Our kitchen's plastic-laminate countertop is developing small scratches. Is there a product that will fill in the small scratches and is color-coded to match the color of the counter?

I don't know of any product that will permanently fill in small scratches and knife cuts in plastic laminate. A temporary solution is to wipe the surface with clear lemon oil. This will alter its appearance so that the scratches are not quite as noticeable. Cleaning the counter removes the lemon oil, so the oil needs to be reapplied periodically.

Regluing a Formica Countertop

Our Formica kitchen countertop is glued onto fir plywood. Over the years, the contact cement has let go in spots. Formica is also glued tightly along the edges. Can a glue be injected between the Formica and plywood to reglue them, even though the old contact cement is still there?

The countertop is probably loose because someone periodically places a hot pot on it. When the temperature goes above 150°F, the contact cement softens and starts to let go. If the contact cement is not too old, it may be reactivated by heating the area with a gun-type hair blower. You need a temperature of about 200°F on the surface of the Formica to reactivate the cement. Then, roll the area with a roller and cover it with a board that is clamped in place.

If tile contact cement is aging and cannot be reactivated, you can bore a tiny hole into the area from the underside of the plywood so as not to damage the surface. Inject a white wood glue into the area. Push the loose section of the countertop up and down to spread out the glue. Then cover the area with a board and clamp it in place until the glue dries.

Plastic Laminate Maintenance

My 30-year-old Formica countertop is in good shape. Although polishing helps to rejuvenate its shine, it makes every fingerprint appear more noticeably than if the laminate had not been polished. What's the best way to maintain the counter's appearance while reducing the tendency of fingerprints to appear as streaks?

Polishing is not required for any laminate surface manufactured by Formica Corporation. All that is required for any of the company's laminates is cleaning with a soft cotton cloth and a mild liquid detergent or household cleaner. The surface should be dried with a clean cotton cloth to avoid spotting. If your counter has a high-gloss finish, residual streaks might remain after cleaning. In this case, the company recommends removing streaks with a mild glass cleaner and a soft cotton cloth.

Refinish a Stove Top

Is it possible to refinish a stove top? I want to make it a different color. If so, what kind of paint would I need?

If the stove is an antique, its top is probably porcelain rather than baked enamel. But regardless of whether the stove is an antique, refinishing its top is not a do-it-yourself project.

Here are a few of the challenges that stove-restoration people have to deal with. For one thing, they start with clean metal. This requires sandblasting. It must be done very carefully so the stove top is not warped in the process. If the top has a baked-enamel finish, restoration companies apply a powder coat to it. As part of this process, they have the means of heating the top to about 400°F. If the top is porcelain, they remove that and then heat the top to 1400°F. A new porcelain coating is then melted onto it.

As you can see, it's not just a simple matter of carefully applying a coat of oil-based paint. You should also consider that on a gas stove, the top could easily reach 250 to 300°F. Furthermore, most commonly available oil-based paints are porous compared to a factory-applied coating, especially porcelain. The combined action of heat and food spills would stain the paint and cause it to turn yellow.

Softener Bypass

We have a water softener in our house. Unfortunately, it adds a lot of salt to our water. For health reasons, we don't want to drink this water because of the high salt content. As a result, we buy bottled water for drinking and cooking. However, we can't use our refrigerator's ice maker because it uses the softened water. Is there a way to bypass this and pipe in unsoftened water?

Cut into the water supply line before the softener and install a T or Y fitting. Run a 1/2-inch-diameter copper pipe from this fitting to the back of the refrigerator. You can then tap into this pipe and run a copper tube to your ice maker.

Reducing a Hood Vent

The previous owner of my house installed a kitchen-range hood vent that exhausts into the attic. I will soon have my roof replaced and, at that time, will run the vent through the roof However, the pipe from the hood into the attic has a 7-inch diameter and I want to reduce this to 4 inches before extending it through the roof. Will such a reduction cause a dangerous buildup of grease in the pipe?

Yes, it could be a problem. The particles of grease that are in suspension will be deposited mainly on the reducing coupling and the walls of the 4-inch-diameter extension. All kitchen range vent pipes should be inspected periodically and cleaned if necessary, but since the reducer and extension would be located in the attic, they would probably be forgotten. The resulting buildup of grease would be a potential fire hazard.

The existing vent pipe should be extended through the roof without reducing its diameter. In addition to being safer, your exhaust fan will operate more efficiently without the added resistance that would result from the reduction. The pipe should be flashed properly at the roof joint, and have a storm collar and a rain hood. Also, if there is no damper over the fan, one should be installed.

Wood-Stove Chimney Connection

I recently bought a wood stove and I'm thinking of using it in the kitchen of our old farmhouse to take advantage of an existing chimney. To

do this, the vent pipe would have to run about 6 inches out the back of the stove, then up 5 feet, and then back another foot or so into the chimney. I'd prefer to cut another hole in the chimney directly behind the stove and run the vent pipe straight in. Which installation is best?**

Before you do anything, have a chimney sweep check the chimney. It's likely the chimney has deteriorated.

With a long vent pipe, there is a large amount of exposed surface area that will radiate heat into the room. However, a wood stove generally produces more heat than is normally needed in a room. Also, the elbow sections will introduce additional resistance to the flow of combustion gases, and they will require frequent cleaning of soot and creosote.

LAUNDRY ROOM

Second-floor Laundry

Our plans for a new Colonial-style home have the laundry room located on the second floor. Can you provide guidelines or tell us of any special requirements?

For special plumbing requirements you should check with your local municipal building department. In all new construction, the department approves plans and inspects the plumbing to ensure that it complies with all applicable state, and local codes.

I have some thoughts on a second-floor laundry that are not code related, but that you should consider. Since there is always the possibility that the washing machine could overflow and flood the floor, I would recommend that during the construction of the house, an auxiliary drain pipe be installed that runs up to the floor of the laundry room. This pipe should be connected to a sheetmetal or plastic pan. The washing machine is installed on top of the pan. Any overflow from the machine will be collected in the pan and drained away, rather than wetting the floor and ceiling below.

I'd also recommend that ball valves be installed rather than globe or gate valves for the hot- and cold-water inlet pipes that supply the washer. Since there is the possibility that the hose connecting the valves to the washing machine could have a defect that causes it to split, you should shut the valves after each use. With a ball valve, you just turn the handle 90 degrees. Other valves require that you turn the handle until the valve is closed.

If you have an electric dryer installed, your electrician should be aware that recent changes to the National Electrical Code require that the dryer be installed with a 4-prong plug and a matching receptacle, not a 3-prong plug and receptacle. That is, the cable supplying the dryer outlet should have two hot wires, a neutral wire and a separate ground. The outlet itself must accept this cable, and the cord from the dryer to the outlet must be a 4-prong type. Finally, the dryer duct should discharge to the outside through a metal, not plastic, vent pipe.

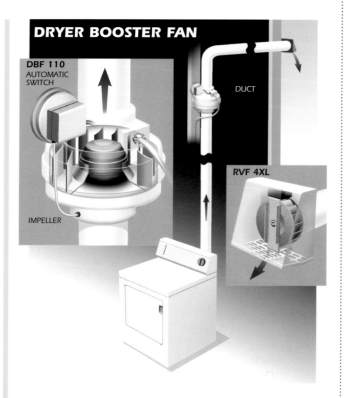

DRYER BOOSTER FAN

DBF 110
AUTOMATIC SWITCH

DUCT

RVF 4XL

IMPELLER

Dryer Venting

My problem concerns the dryer vent in our 12-year-old house. The vent extends through the roof, where it is capped by a screen that fills with lint. I have to climb up on the roof to clean it out. I asked the company that installed the dryer about running the vent horizontally to the nearest outside wall. They said it was too long a distance, and suggested I install a device to trap the lint in water. Do you have any suggestions?

My first suggestion is not to listen to the company. Trapping the lint in water increases humidity in the house and adds resistance to the dryer's exhaust. According to building codes, the maximum length for a straight dryer duct is 25 feet. For every 45-degree bend in the duct, its length should be reduced by 2

feet, and for every 90-degree bend, it should be reduced by 5 feet. Bends in the duct increase its resistance to airflow, so its length has to be shortened accordingly. If duct resistance is too great, it can overcome the dryer's ability to exhaust lint and humidity. The lint will settle in the duct, and the humidity in the exhaust can condense. The wet lint can partially or fully block the dryer duct, and this can cause the dryer to run longer than it would otherwise. Aside from wasting energy, the accumulated lint will dry out and become a fire hazard.

You can prevent this problem by installing a booster fan in the dryer duct. The fan must be installed at least 15 feet from the dryer's back. This will allow the lint to dry before it reaches the fan. Wet lint will clog the fan, but dry lint will pass through it. As a precautionary measure the fan casing should be opened and inspected once a year. Check with the manufacturer to determine the maximum duct length that can be used with the fan.

Venting Gas Dryer

I'm having a gas dryer installed in the basement of my home. However, I am having trouble venting the dryer. It seems impossible to run the vent through the house's thick stone foundation, and it can't be vented through the sides of the house, because of attached houses on both sides. The plumber installing the appliance advised venting it into the boiler flue by using a special fitting. Is this installation safe and legal? Your recommendation will be appreciated.

The primary purpose of a dryer vent is to discharge the moisture-laden air and lint that results from drying the laundry. This holds true whether the dryer heats its air stream with a gas flame or an electric heating element. The only difference between the two is that the air stream from a gas dryer will contain small amounts of combustion gases which should be vented outside. Both electric and gas dryers should be vented outdoors through a dedicated metal vent. In either case, the vent must not be connected to your boiler flue. It is illegal to vent a dryer through a flue for a boiler or furnace because moisture and lint from the dryer will eventually damage the flue and block it, causing combustion gases from the boiler or furnace to spill back into the house.

At one time, dryer manufacturers sold lint-catcher kits to be attached to dryer vents for indoor venting. They no longer do so, and now recommend venting a dryer to the outside through a dedicated metal vent. Several problems were associated with indoor venting, and among them was the introduction of moisture into the house. Also, people did not clean the lint catcher frequently enough. A lint buildup is a fire hazard and causes the dryer to work inefficiently. Check and clean a dryer vent once a year.

Although it may be difficult, a hole can be made through the rear foundation wall to vent the dryer. You may need the assistance of a mason.

DAMPNESS, HUMIDITY

Severe Humidity

I have a severe humidity problem in my house, which is two years old. The indoor relative humidity levels run 70 to 80 percent during the winter months. My double-glazed windows are constantly covered with moisture and the varnish is beginning to peel. The attic is vented by six roof vents and six eave vents. The walls contain 6 inches of Fiberglas insulation with plastic vapor barrier and there's 12 inches of cellulose insulation in the attic. To take care of moisture in the bathroom, I have a power vent to the outside. The range hood is not vented but there is only minimal cooking done on the range.

It sounds as if you have a tight, energy-efficient house. While this helps keep your heating costs in line, it inhibits fresh-air infiltration that would otherwise help to control humidity. Condensation problems often appear in the fall after the house has been exposed to the warmer moist air of the summer months. Everything in the house including wood framing, walls, floors, clothes, furniture and so on absorbs the moisture from summer air.

When the drier, cooler weather arrives, the house tends to dry out and thereby increases the relative humidity of the interior air. If the temperature at the windows is below the dew point, this moisture will condense.

If this is the cause for the moisture buildup in your house, it can be corrected by ventilating the rooms. Open the windows in each room for about an hour during the warmest part of the day. The dry outside air can then mix with moisture-laden household air and reduce the overall humidity.

There are other possible causes for moisture buildup in your house. If your basement takes on water after a rain, make sure it's pumped out as soon as possible. The water vapor will migrate up to the habitable portions of the house. Try to avoid storing firewood in your basement. As the wood dries, the moisture it loses will contribute to the relative humidity of the household air. Also, if there is a dirt floor in the crawlspace it should be covered with 4- to 6-mil polyethylene sheets even if the soil feels dry. According to the Small Homes Council at the University of Illinois, a 1000-square-foot house can release as much as 18 gallons of water per day through evaporation in the crawl space.

Another possible source for excessive moisture buildup in the house is a clogged heating-system chimney. One of the products of combustion in a gas-fired heating system is water vapor. If the chimney is clogged, the water vapor will pour into the house. This must be checked not only because of the moisture-level buildup, but because poisonous carbon monoxide could be leaking into the house as well.

Sweaty Windows

The windows in my house sweat and produce excessive moisture. Water accumulates on the window sill and literally needs to be dried up with a sponge or rag. This occurs daily in the wintertime. The house is three years old, and water damage to the drywall is already visible. Any suggestions to eliminate the moisture buildup would be appreciated.

The condition is obviously the condensation of excessive moisture present in your house. Although I can't determine the exact cause of the problem, I can recommend three possible sources for excessive moisture.

Many homes with warm-air heat have a humidifier mounted on the furnace. If you have one, the unit may be malfunctioning, introducing too much moisture into the airstream.

Another moisture source is a bathroom exhaust fan that vents into the attic, and not outside. The fan may also be blocked, or its outlet in the attic floor may be covered by insulation. The fan should exhaust moisture into a duct that leads to a roof vent or through a sidewall with a rainhood for protection.

A third source is a dirt crawlspace under the house. Moisture from the soil is pulled into the crawlspace, then into the house. In this case, install a vapor barrier of plastic sheeting above the dirt floor and ventilate the crawlspace. There should be at least two vents, with a total free area of $\frac{1}{1500}$ of the crawlspace area.

Condo Humidity

I need advice on how to deal with a humidity problem. I close my condominium apartment for a few weeks at a time. When I return, I find a bad case of mold and mildew I'm considering getting an electric dehumidifier to solve the problem. Is there a better solution?

Whether or not a single dehumidifier will be adequate for your purpose depends on the configuration of your apartment.

If you have a number of rooms, you'll probably need more than one dehumidifier because the partitions will interrupt air circulation between the rooms. You can substitute a fan for one of the dehumidifiers if it's positioned so that air is moved from a moisture-laden room to one with a dehumidifier.

Dehumidifiers are usually equipped with an overflow control that shuts the unit off when the reservoir is filled. To make sure that the units keep working, run a hose from each one to a sink or toilet.

If your condominium apartment is centrally air conditioned, then a good cost-effective solution is to install a humidistat in parallel with the air conditioner thermostat. With this system, whenever the humidity builds up beyond a preset limit, the air conditioner will be activated even if the temperature in the apartment is not high. This will remove excess moisture from the air throughout the apartment.

Condensation in Sunroom

During cold weather, condensation forms on the windows in my unheated sunroom adjacent to the living room. A pair of French doors provides access to the sunroom which, incidentally, was built over a concrete slab. The condensation is so heavy that it forms mildew on the walls. A dehumidifier has proven itself ineffective. How can this be corrected?

Although the sunroom is closed off from the living room, it's likely that water vapor is migrating from the living room into the sunroom. Moisture will travel from a room with high vapor pressure to one with lower pressure. In most cases, vapor pressure is higher in a warm room than in a cool room. Check the wall around the French doors and seal any openings you happen to find. Seal the perimeter of the doors with weatherstripping. You can also tape a large polyethylene sheet over the doors and their molding to see if this decreases the condensation. The plastic needs to be taped over the doors on the warm side of the room (the living-room side).

If the problem persists, I would suspect that water vapor is entering the room through the concrete floor slab. When the room was built, a plastic-sheet vapor barrier should have been installed under the slab. You can determine if moisture vapor is migrating through the slab using a simple test. Place an 8-inch square of aluminum foil on the floor, and tape down its perimeter. If, after a day or two, the underside of the foil is damp, you've located the source of the problem. In this case, you can use an electric heater and warm the sunroom to reduce or eliminate the condensation. Heat helps the moisture vapor remain suspended in the air, and it warms cold surfaces, which further decreases the chance of condensation.

FIRE SAFETY

Fire Extinguishers

I want to buy a fire extinguisher for my home, but I don't know which type to get.

This is a good question because using the wrong type of fire extinguisher could do more harm than good. It must suit the type of fire that's burning.

There are three types of fires. Class A fires are those involving ordinary combustibles such as wood, paper, cloth, rubber, and so on. Home fires of this type often start in the living room or bedrooms. Class B fires involve cooking oils, grease, gasoline, paint thinners, and other flammable liquids. These fires generally break out in kitchens and garages. Class C fires are electrical fires, and are usually the result of faulty wiring, overloaded circuits, short circuits, or faulty electrical appliances.

On fire extinguishers, the letters A, B, and C within a triangle, square, and circle designate these categories, respectively. Class B- or BC-rated extinguishers are not effective on a Class A fire. Also, water, which is effective in extinguishing a Class A fire, will cause a Class B fire to spread and can cause a severe shock in a Class C fire. Once a fire in the home spreads, it can quickly include all three categories. Therefore, your best choice is a fire extinguisher rated for all three classes of fire.

Fire extinguishers are available at hardware stores and home centers. When you buy one, check to see that it's listed by Underwriters Laboratories, Inc., and displays the A, B, and C designations. Also, note the numbers in front of the A and B designations. These refer to the size of fire that the extinguisher can generally handle. The numbers are not absolute figures, but are relative terms for comparing different units. For example, an extinguisher with a rating of 2A:40B:C will handle a Class A fire twice as large, and a Class B fire four times as large, as a unit rated 1A:10B:C.

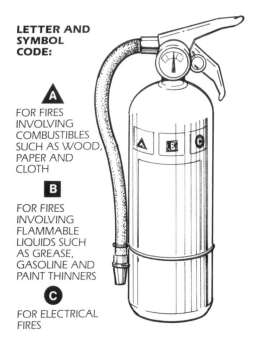

LETTER AND SYMBOL CODE:

A

FOR FIRES INVOLVING COMBUSTIBLES SUCH AS WOOD, PAPER AND CLOTH

B

FOR FIRES INVOLVING FLAMMABLE LIQUIDS SUCH AS GREASE, GASOLINE AND PAINT THINNERS

C

FOR ELECTRICAL FIRES

Note that there are no size ratings for Class C fires. The C designation only means that the chemical inside will not conduct electricity.

Wood-framing Clearance Requirements

My house has a heating system with a metal chimney, which runs up through the attic and terminates above the roof. Since the temperature of the chimney in the attic is not high enough to ignite wood, why is it necessary to have a 2-inch clearance between the chimney and the wood framing in the attic?

The ignition temperature of wood is defined as the temperature at which it begins to burn. Wood and other combustibles undergo a physical change when continually exposed to elevated temperatures. This reduces their ignition temperatures.

Wood normally begins to burn at about 400° to 600°F. However, when it's continually exposed to temperatures between 150° and 250°F, its ignition temperature can be as low as 200°F.

The lowering of the ignition temperature of wood and other combustibles can take years to occur. When it does, should the wood's temperature coincide with its lowered ignition temperature, it will ignite and burn spontaneously. This is the basis for the clearance requirements in codes.

Fire-extinguisher Service Life

For 35 years I have been very concerned regarding fire protection and therefore I have always kept a variety of fire extinguishers at hand. I regularly check their condition and pressure gauges. Some of the fire extinguishers are 20 years old and have never been used. When is it advisable to service or discard a fire extinguisher, even though the gauge indicates it's okay?

Even though the gauge on your old fire extinguisher indicates that the unit is fully charged, you don't want to find out during a fire that it's incapable of putting out a fire. Your safest bet, according to Kidde, a manufacturer of fire extinguishers, is to follow the recommendations of the National Fire Protection Association (NFPA).

Residential fire extinguishers come in two types—rechargeable and nonrechargeable. The NFPA recommends that after six years rechargeable extinguishers that use dry chemicals be inspected. Rechargeable extinguishers that use water, carbon dioxide, or foam should be inspected every five years. In both cases, a fire-extinguisher service company should perform the inspection and recharging operations. The inspection should include disassembly and examination of the internal cylinder surface as well as all components. The extinguisher should then be refilled with the appropriate

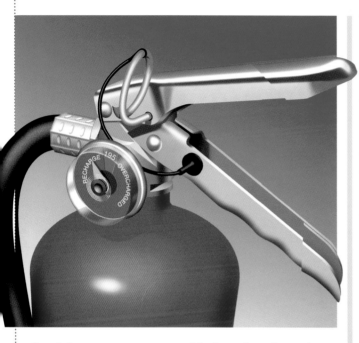

fire-fighting agent, reassembled, and recharged. It's likely this will cost more than a new unit, which in my opinion is the better option.

Nonrechargeable extinguishers do not require the six-year internal inspection. However, even if the gauge on the unit indicates that the unit is charged, the extinguisher should be removed from service 12 years after its date of manufacture.

In addition, all extinguishers should be periodically inspected to see that the unit is unobstructed, that the seal or tamper indicator is not broken, and that there is no evidence of corrosion or other outwardly evident damage. The pressure gauge should indicate that pressure is in the operating range.

Kidde further recommends that extinguishers be inspected monthly for insect infestation. Although this does not seem likely, insects have been known to build nests in extinguisher valves, preventing their proper operation.

Chirping Smoke Detector

I have a problem with my smoke detector chirping. The batteries have been changed, but it doesn't seem to help. What is the problem and how can I fix it?

As you know, a smoke detector chirping generally indicates that the battery is weak and should be replaced. It will also occur if the battery is not properly installed. Check that the battery is snapped firmly into position. Dirt or dust on the alarm cover can cause chirping also. You can clean the cover by gently brushing a vacuum over it.

If this does not correct the problem, then there is a malfunction in the detector and it should be replaced. If the device is still under warranty, contact the manufacturer for a replacement.

Do you have other smoke detectors throughout your home? How old are they? Fire-safety professionals recommend replacing smoke alarms before they are 10 years old. If you are not sure how old they are, play it safe and replace them.

Smoke Detector Location

To afford the earliest possible warning in a fire, it's important that a smoke detector be located properly on a wall or ceiling.

Smoke is lighter than air and rises with convection currents, but not as a stratified layer. Smoke will bounce off the walls as it rises, leaving a dead airspace about 4 inches wide and deep at the corner of the walls and ceiling. A smoke detector placed in the space may not detect smoke until the room is filled

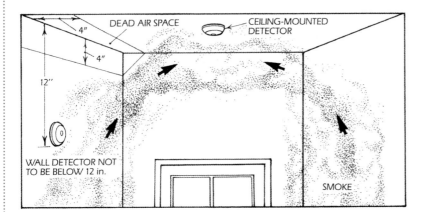

with smoke, precluding the possibility of an early warning and a safe escape.

Ceiling-mounted smoke detectors should be mounted outside the dead airspace. Wall-mounted smoke detectors should be mounted below it, but not more than 12 inches below the ceiling. Neither wall- nor ceiling-mounted detectors should be placed near a light fixture or a ventilation grille that could block smoke from reaching the detector.

Smoke Alarm Signal

My husband and I were awakened recently when all our hardwired smoke detectors went off at once. We checked everything and found no problem. Do you have any idea what causes this?

A smoke alarm can be triggered by a buildup of dust, dirt, or an insect building a nest inside it. To prevent this, gently vacuum a smoke alarm periodically, and change its battery once a year. When smoke alarms are hardwired and one unit is activated, all the others will be activated. The fact that the alarms went off at the same time indicates that they are working properly.

Home Fire Sprinklers

Affordable and dependable sprinkler systems with low-profile heads are available for one- and two-family homes. A sprinkler system combined with smoke detectors provide the most reliable form of residential fire protection for you and your family. Excluding deaths by explosion or flash-fire, there are no known cases of multiple deaths in a fully sprinklered building due to fire or smoke, according to the U.S. Fire Administration. Though 80 percent of fire deaths occur in residential buildings, they rarely have sprinklers.

In 1976, less than 5 percent of U.S. homes had a smoke detector. Today, most homes have one or more.

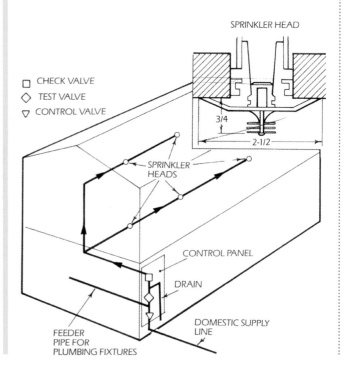

Smoke detectors make a vital contribution to fire protection by providing early warning, but sometimes even this is not enough. About 50 percent of the fire death victims were very young or old, handicapped or intoxicated, and were unable to escape.

Residential sprinklers are designed to protect people in the area of fire origin and reduce fire damage. They activate five to seven times faster than standard commercial sprinklers and have a different spray pattern and droplet size. Their quick response keeps toxic gases and smoke to a minimum.

Also, they prevent fire from growing to the flash-over stage. This occurs when gases from burning materials accumulate and explode. Generally, flash-over takes place when temperature at ceiling level reaches 1000° to 2000° F. With many combustible materials being introduced into homes in construction or as furnishings, the flashpoint can be reached in two to three minutes.

Designers paid special attention to the appearance of the sprinkler head. It projects only about ¾ in. from the finished ceiling. Many use escutcheon plates that can be painted or plated. The heads thread into feed pipes.

Flame Spread

Can you tell me what "Flame Spread 200 or Less" means? I saw it on a label on the back of a 4 x 8 plywood panel that had a decorative finish.

Flame spread is the propagation of a flame over a surface. The flame spread rating classifies the fire hazard potential of different building materials. It is based on tests performed by independent labs using procedures developed by Underwriters Laboratories. Inorganic materials and untreated red oak provide the range against which flame-spread ratings are made. Inorganic materials, which include concrete, cement-asbestos board and metal, have a flame spread of 0. Untreated red oak has a flame spread of 100.

Building codes require materials that have a low flame-spread rating (0-25) to be used in fire escape routes, such as in stairways and exits. Where passageways and corridors are not part of an enclosed exit, the flame-spread rating of the material should not exceed 75. Materials used for interior walls and ceilings generally have a flame-spread classification of 200 or less. This includes most untreated plywood or paneling.

CEILING FANS

Ceiling-fan Blades

I am thinking about installing ceiling fans in my new condominium. Is a fan with five blades more effective than a fan with just three or four blades?

The number of blades on a ceiling fan is more a matter of personal preference than effectiveness. A fan with three blades can be just as effective as a fan with five. Effective airflow from a ceiling fan is a function of several factors: motor speed, blade pitch, blade width and blade diameter. Also, it should be installed in areas where it can move the most amount of air. One company that manufactures fans suggests the following fan sizes:

- 36-inch fan for areas up to 50 square feet such as bathrooms and hallways.
- 42 inch fan for areas up to 75 square feet such as breakfast nooks and laundry rooms.
- 50- to 52-inch fan for areas up to 100 square feet, such as small bedrooms and small kitchens.
- 54-inch fan for areas up to 400 square feet, such as standard-size bedrooms and family rooms.

Wobbling Ceiling Fan

I recently installed three new 42-inch ceiling fans in my house. All three have light fixtures on them. Two of them run evenly and smoothly. The third one vibrates and shakes. Can anything be done to make the fan run smoothly?

There are several reasons why a ceiling fan wobbles when operating. The fan blades could be out of balance, out of track, or warped. The wobble could also be caused by the air turbulence that results from the blades being less than 6 inches from one side of a sloped cathedral ceiling or ceiling beam.

You can check to see if the fan blades are out of balance by clipping a spring-type clothespin halfway between the tip and the blade iron on the leading edge of one blade. Turn on the fan to low speed and see if the weight stops the wobble. Try each blade to determine if it needs more weight. If the clothespin stops the wobble, it should be replaced with thin adhesive-backed, lead-weighted tape, which is available through the fan manufacturer's service center.

Check to see if one or more blades are out of track. Using a yardstick, measure the distance from the ceiling to the tip of each blade. The distance should be equal for each blade. If it isn't, contact the nearest manufacturer's service center about correction or warranty replacement. If the blades are not out of track, try swapping blades to correct the problem. Switch the position of two adjacent blades while leaving the other two in their original positions. While the blades are off the fan, lay them on a flat surface to see if they are warped. If so, replace the blades.

Eliminating Ceiling-fan Wobble

I have a 52-inch, four-blade ceiling fan that wobbles badly. The ceiling box that supports the fan is very solid. It is a ball type mounting. My guess is that the blades are not balanced. Can you give me a methodical way of balancing the blades? The manufacturer did not number the blades in sequence for mounting.

A wobble in a ceiling fan usually is caused by blades being out of balance or out of track. However, it can also be caused by the blades being closer than 6 inches from one side of a sloped cathedral ceiling or a ceiling beam.

If this isn't the case, check to see if one or more of the blades is out of track. Using a yardstick, measure the distance from the ceiling to the tip of each blade. It should be the same. If it isn't, call or contact the nearest manufacturer's service center about correction or warranty replacement. If the blades are not out of track, try swapping blades. Switch the position of two adjacent blades while leaving the other two in their original position. While the blades are off the fan, lay them on a flat surface to see if they are warped. If so, they will have to be replaced.

If you still have a wobble after swapping all the blades, try clipping a spring-type clothespin halfway between the tip and the blade iron on the leading edge of one blade. Turn the fan on low and see if the weight stops the wobble. Try each blade to determine if it needs more weight. If the clothespin stops the wobble, it should be replaced with thin adhesive-backed, lead-weighted tape, which is available through the manufacturer's service center.

Balancing a Ceiling Fan

Could you please tell me an easy way to balance the blades on a ceiling fan? I have two fans that vibrate, and would like to fix them.

A number of factors besides improper balance will cause a ceiling fan to shimmy or wobble. Check to see if one or more of the blades is out of track. Using a yardstick, measure the distance from the ceiling to the tip of each blade. It should be the same. If it isn't, call or contact the manufacturer's nearest service center about correction or warranty replacement. If the blades are not out of track, try swapping blades. Switch the positions of two adjacent blades while leaving the other two

in their original positions. While the blades are off the fan, lay them on a flat surface to see if they are warped. If so, they will have to be replaced.

If the fan still wobbles after swapping all of the blades, try clipping a spring-type clothespin halfway between the tip and the blade iron on the leading edge of one blade. Turn the fan on low and see if the weight stops the wobble. Try each blade to determine if it needs more weight. If the clothespin stops the wobble, replace it with thin, adhesive-backed lead tape, which should be available through the manufacturer's service center.

Ceiling-fan Direction

We have a two-story contemporary home with a ceiling fan downstairs in the living room and another upstairs in the loft area. In the summer, we open the windows in the loft above the fan and run both fans so they blow air up and out the windows. In the winter, we run the fans down so they blow the warm air from the second floor down to the first floor. Recently I was told this is not the correct way to use the fans. Is it?

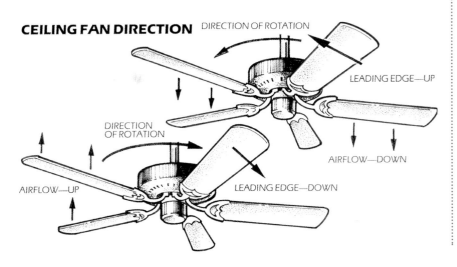

CEILING FAN DIRECTION

DIRECTION OF ROTATION

LEADING EDGE—UP

AIRFLOW—DOWN

LEADING EDGE—DOWN

DIRECTION OF ROTATION

AIRFLOW—UP

A ceiling fan is not intended to be used as a whole-house fan, which is how you are using it during the summer. Whole-house fans are very effective, and if that's what you want, you should use a fan that is designed for that purpose.

Ceiling fans are designed to circulate air in a "closed" environment. A ceiling fan should force air down during the summer. This downward airflow cools your skin as it moves over it. During the winter months, the fan should rotate so it produces an upward airflow.

During the winter, heated air rises toward the ceiling, and cool air settles toward the floor. Depending on the room size and shape, there could be a 15°F difference between the floor and ceiling. Intuitively, people want the fan to pull warmer ceiling air down during the winter and vice versa during the summer.

The fan speed should be fast enough to break up stagnant air trapped in the corners and in the peaks of sloped and cathedral ceilings, but slow enough so it doesn't create a draft.

An easy way to determine whether the fan airflow is up or down is to look at the fan blades as they rotate. The blades are installed on a slight angle. If the leading edge of the blade (the edge facing the direction of rotation) is up, the airflow will be down. And when the leading edge is down, the airflow will be up.

MISCELLANEOUS

Mirrors Mold Buildup

My decorative glass mirror tiles have developed dark, moldlike, ugly growths in the backside reflective films. The tiles are the inexpensive type with stick-on tapes. I intend to replace the damaged tiles, and wonder whether first painting the tile backs or applying silicone caulking would keep the problem from recurring.

The dark blotchy areas on your mirror tiles are not the result of mold buildup, but rather the tarnishing (oxidation) of the silver backing. Mirrors are made by covering large pieces of plate glass with a film of silver, and then a coat of paint to protect the silver from scratches and tarnishing. These large mirrors are then cut into the desired sizes. The cutting not only fractures the glass but affects the paint bond at the cut line. Minute amounts of the silver film are then exposed to the air where the silver can tarnish. Therefore, additional paint could help if you pay particular attention to the tile edges.

Proper handling of the mirror is also important. If, when being installed, the backing is scratched, even if the scratch is tiny and doesn't penetrate the silver, that area will eventually tarnish.

I don't recommend silicone caulking for your mirror tiles. It has a solvent that will attack the mirror backing. You should use an asphalt-base adhesive (mirror mastic) or foam adhesive tape, both of which are available at hardware stores.

Tempered Glass

How can I cut tempered glass to the size that I want?

You can't cut tempered glass. Used as a safety glass, it is 4- to 5-times stronger than ordinary plate glass of the same thickness. It cannot be cut or drilled after tempering. If you try to cut tempered glass, it will break into small, cube like fragments, unlike plate glass, which breaks into large shards.

Static Electricity: A Shocking Problem

Static electricity is a problem at our house. My wife tries to use additives in the washing machine and dryer to control it, but it doesn't help. Every time I put on a sweater and take it off—especially the latter—I can hear crackling. The problem is the worst in the bedroom upstairs. Just walking across the carpeting, I can pick up static electricity in my socks. It even seems to be in the air, floating or hanging there. What can be done about this?

Static electricity in the house is usually more of a problem in the winter, when the relative humidity is very low, than in the summer when the humidity is high. This is a clue to how you can minimize the problem.

Static electricity is the buildup of an electrical charge brought about by rubbing two dissimilar nonconducting materials together. Moist air is a better conductor than dry air and, as such, helps dissipate the charge before it becomes noticeable.

According to a Scandinavian new-technology magazine in an article on Danish heating, ventilating, air conditioning and energy technology, studies have shown that in order to prevent static shock in rooms with carpets of wool, nylon, and some other synthetic fibers, the relative humidity should be higher than 40 or 50 percent. For carpets of cotton and sisal, the relative humidity should be higher than 30 or 35 percent.

Some manufacturers have introduced conducting fibers in the carpet to minimize the problem. In addition, there are products on the market that can be sprayed onto a sweater or other article of clothing to eliminate static cling.

Low Phone Volume

Why is it that our transmitting and receiving volume goes down in proportion to the number of phones that are in use? We have observed this in new houses in the United States and where we live now, in Korea. What can be done to remedy this?

In a typical residence, there is a fixed power input for the telephone signal. This signal has enough strength to activate the small speaker in the handset. However, when one or more telephone extensions are used to join in the conversation, resistance is added to the circuit. The signal to the initial phone is weakened because part of the signal power is diverted to each extension, and the sound volume is reduced proportionately.

You can have a computerized control installed that amplifies and balances the signal to provide equal sound volume on all extensions. However, this system can cost about $1000. Another approach is to buy handsets with built-in amplifiers. Compared to a computerized control, these are inexpensive—about $100 apiece. On the other hand, they amplify background noise too, and this can be annoying.

Water-based Versus Oil-based Paint

Can I apply a water-based paint directly over an oil-based paint? And conversely, can I apply oil-based paint over a water-based paint? I have had conflicting advice from paint stores.

You can paint over a water-based paint with oil-based paint, and vice versa. However, you should not paint over a very smooth or high-gloss surface with either product without first roughing the surface. If the initial surface coat is flat, clean, and free of chalklike powder, there is no need to use a primer.

On the other hand, if the initial surface has a gloss, and you sand it to rough it up, you should use a primer. The primer will ensure a good bond to the surface beneath and will help produce an even level of gloss on the finish coat.

Central Vacuum System

We are wondering if it is possible to use a shop vacuum for a central vacuum system. Are there instructions on using a shop vacuum in such a way?

I have never seen instructions on how to use a shop vacuum in a central vacuum system, and I suspect that retrofitting one for this purpose is so involved that it's not worth the effort.

There are two problems that have to be addressed in installing a central vacuum system. Normally, there is a long run of duct between the vacuum unit and the various wall outlets. The greater the distance, the greater the resistance to airflow. To overcome the resistance, a vacuum unit that produces a powerful suction is needed. A shop vacuum is generally not as powerful as a central unit, and as a result you are likely to get minimal or inadequate suction at the remote outlets.

The second problem is that, in a central system, there is a low-voltage microswitch in each wall outlet that activates the vacuum motor, which operates at 120 volts. Consequently, a relay and associated circuitry are needed so that each outlet can activate the vacuum motor. You would have to find the appropriate components and retrofit them to the shop vacuum.

Cutting Ceramic Tile

Occasionally a homeowner is faced with replacing a portion of a ceramic tile rather than the whole tile. Can the tile be cut by a homeowner? If so, how would he go about it?

Yes, a tile can be cut by a homeowner, but the job requires a special tool. Since a tile is usually $5/16$ to $3/8$ inch thick, you cannot simply score the surface and snap it at the edge of a tabletop. You can buy a variety of tile-cutting tools at a home center, ranging from small wet-cutting circular saws to score-and-snap devices, but the cost may not be justified for a small job.

There are two approaches. If you only have a few tiles to cut, mark them and bring them to a tile store. Ask them if, for a few dollars, they would be willing to cut the tiles. The other option would be to rent a tile cutter from a tile store. The rental cost for one of these machines is only about $10 a day.

When removing the tile to be replaced, you should be careful not to damage the adjacent tiles. First, use a grout saw to remove the grout around the tile. This is not a mechanical saw, but a simple abrasive tool that you run along the grout lines. After the grout is removed, use a cold chisel and carefully chip away the old tile. Begin the process by chipping away the tile's center, then work to its edge. Wear safety glasses when doing this work.

Radio Frequency Interference

When my Honeywell electrostatic air cleaner is running, it produces a snowy condition on my television. What causes this and how can I correct it?

You're getting radio frequency interference (RFI) from the air cleaner's high-voltage power supply. Radio waves radiate outward in all directions and can be picked up by a household electrical wire, which acts as an antenna. A lack of electronic shielding on the air cleaner or the television or an improper ground (at the air cleaner, television or the house's service panel) contributes to the problem.

Have an electrician evaluate the quality of the ground at the house's electrical supply and at the supply voltage to the air cleaner. If electrical grounding is not the problem, try installing an RFI filter on the line coming into the air cleaner.

Stairway Design

I am planning to build an addition to my house, and I want to build an interior stairway for access. Are there specific design dimensions that I should follow from a safety point of view?

The two main concerns in stairway design are the treads (the steps) and the risers (the vertical sections that link the steps). It's important that the riser height be the same for all the steps. Otherwise, the dimensional variation interrupts the natural rhythm of ascending and descending the steps, producing a tripping hazard. Also, the treads must not be too narrow, otherwise, a person would need to place his foot diagonally on the treads for safe and comfortable support.

The U.S. Department of Housing and Urban Development (HUD), in its "Design Guide for Home Safety," recommends the following design specifications for interior stairs. The maximum riser height should be 7½ inches. Minimum run width should be

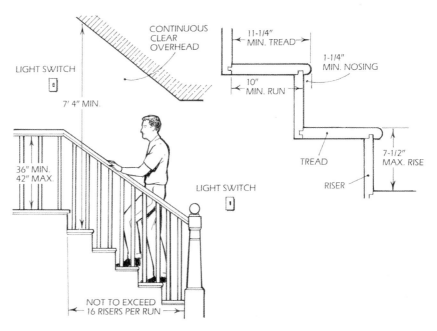

10 inches, and minimum tread width, 11¼ inches. The maximum nosing width should be 1¼ inches.

Also note the position of the handrail and light switches on the drawing and the ceiling height requirements. Maintaining adequate ceiling height is important to avoid low headroom at the bottom of the stairs.

Finally, check with your local building department for further requirements.

Metal Roof Condensation

I built a porch with a metal roof over it. We closed it in using storm windows, but have not installed heat. In the winter, condensation forms and runs off the inside of the roof. What is the best way to stop the condensation? Will installing plywood and shingles over the metal help?

The way to stop the condensation from developing is to prevent the vapor in the porch air from contacting the cold metal roof. Putting plywood and shingles on top of the roof will not help.

Instead, you should put insulation underneath the roof and cover the insulation with a vapor barrier. With this installation, the vapor will not migrate beyond the vapor barrier. Also, because of the insulation, the vapor barrier will have a temperature that is above the dew point. Consequently, condensation will not develop. You can cover the insulation and vapor barrier with drywall or wood.

10. Basement and Crawlspace

WATER SEEPAGE ● SUMP PIT/PUMP ● DAMPNESS/ CONDENSATION ● FOUNDATION ● CRAWLSPACE

WATER SEEPAGE

Basement Water Seepage

We have a paneled, finished and furnished basement that's being destroyed by water seepage, and I'd like to know which kind of contractor to call to fix the problem. I've spoken to a landscape contractor who said that dry wells will direct water away from the house. Will this correct the problem? Or should I call a general contractor for foundation work? I recently had the wall treated by a waterproofing company, but it didn't do any good.

The contractor you should consider hiring to correct your problem is one who specializes in waterproofing basements. You don't need a general contractor, because that person would hire a subcontractor to do the job—and you'll be paying for the services of a middleman. But before hiring a contractor, there are a few important things you should know about water seepage into a basement.

Water can seep into a basement through the foundation walls, through the floor and through the joint between the floor and the walls. In order to correct the problem, it's important for the homeowner to determine the location of the water entry. For example, if water is seeping in through the floor or floor joints, then waterproofing the walls won't correct the problem. Similarly, if water is seeping in through the walls, then installing a waterproofing system in the floor won't help.

Installing dry wells won't correct your problem unless they're tied to drain tiles. Also, if your house is in an area that has a high water table, the dry wells may not be effective during those parts of the year when the water table rises.

Based on your description of the problem, you'll probably need either an exterior or an interior drainage system. An exterior system should have a free-flowing outlet for the drainage piping. An interior system would require a sump pit and pump.

Preventing Water Seepage

My daughter's house has a basement that gets water in it after a heavy rain. Who do you contact to get help? Someone suggested that a mason could put a small trench around the foundation perimeter and install a sump pump.

Before contacting anyone to have a foundation drain installed, check for a number of problems, such as clogged gutters. An overflowing gutter can cause water to seep into the basement. Rainwater should flow freely from the gutters to the downspouts, and the downspouts should direct water away from the house. Also, the ground immediately adjacent to the foundation should slope away from it, not toward it. This prevents water from accumulating around the foundation.

If problems with gutters and downspouts are not causing the seepage, then it's best to intercept the water before it enters the foundation. If the ground slopes toward the house, consider having a curtain drain installed in the sloped area. The drain should have a free-flowing outlet that discharges away from the house.

If sloped ground is not the problem, then water-proofing the house's foundation from the outside is your next, and best, option. However, in existing homes, the cost of temporarily relocating foundation plantings is often prohibitive, so many homeowners opt for an interior perimeter drainage system and a sump pump. In this case, the water is pumped from the pit to an outlet that is well away from the house.

A variety of contractors install interior drainage systems. One way to locate an appropriate contractor is to check your Yellow Pages for information under the heading, "Basements, Waterproofing" or "Foundation, Waterproofing."

Baseboard Drainage System

Water leaks between the foundation wall and my basement floor. I have heard of plastic baseboards that divert this water to a sump pump. Where can I get more information on these products?

The joint between the foundation wall and the floor is a common entry point for water. Ideally, the way to correct this is to intercept or divert subsurface water before it reaches the foundation. However, if this is not practical, plastic baseboards can control the seepage.

The baseboards can be used with either poured concrete or concrete-block foundation walls. They consist of preformed hollow plastic baseboard sections, splices, inside and outside corners and endcaps. The system is secured to the concrete floor slab, adjacent to the wall joint, with an adhesive sealant that cures even under wet conditions. They are sold through hardware stores, and home centers.

Leaking Basement Wall

I have a nagging problem with water seeping into my basement after a heavy rainfall. Perforated drain pipes under the basement floor do a good job of relieving hydrostatic pressure below the slab. However, the cavities in the concrete block walls fill with water, which leaks into the basement.

Contractors recommended breaking up the floor at the perimeter and installing drain tiles and a sump pump. However, I recently heard about a plastic molding, similar to a baseboard radiator-cover, that is sealed to

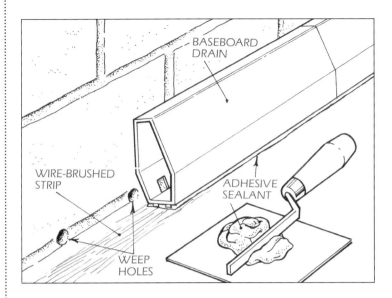

Cellar Leakage

I have leakage in my cellar where the concrete-block walls meet the cellar floor. Occasionally, water will seep through the cellar floor near the block walls. I've received several quotes on waterproofing, and they all use clay or bentonite injected into the ground around the exterior of the house. Some of these companies will guarantee the work they do. Is this a good procedure to follow? I hope to avoid excavating outside the foundation. Thank you for any information you can give me.

the inside of the foundation wall along the floor. Holes are drilled into the blocks to drain the water from the cavities into the molding, which carries it to a sump. Can you tell me more about this arrangement?

There is such a product, and it does work. But the ideal way to correct a hydrostatic leakage problem in a concrete block wall is to intercept or divert water before it reaches the foundation.

However, if you can't lay drainage pipes because the area around the foundation is paved or covered with expensive shrubbery, or is otherwise inaccessible, then you must control the seepage into the basement. You can do this as you describe, by puncturing weep holes at the base of the concrete blocks and directing the effluent to a sump pit, where it can be pumped from the house.

The hollow vinyl baseboard units that collect wall seepage and channel it to a sump are sold at home centers and building supply outlets. Corner sections, splices and end caps are also available. In addition, you will need a special adhesive sealant that cures even under wet conditions. It should be noted that this system isn't viable in cases where water is forced up directly through the floor by hydrostatic pressure.

I wouldn't recommend soil injection with a "waterproofing" material as a means of coating the foundation walls to control a water-seepage condition. To the best of my knowledge, it rarely works over the long term. Unless the foundation is exposed, you wouldn't know if there's anything in the ground preventing full coverage of the wall. For instance, there may be wood debris that was buried by the builder that's now covering a portion of the wall. This debris would keep the waterproofing

material from reaching the exterior foundation surface. Over the years, the wood will rot, exposing a part of the foundation wall through which water can flow. Also, if water is seeping into the house through the floor, waterproofing the walls will not correct the condition.

Since you don't want to resolve this problem by excavating on the outside, adjacent to the foundation, you can do it from within your cellar—provided the floor slab isn't integral with the foundation footings. It'll involve cutting 1-foot-deep openings within the floor slab adjacent to the problem walls, puncturing holes in the bottom blocks, installing 4-inch-diameter perforated drainpipes and gravel and then pouring concrete to match the existing floor.

The drainpipe can discharge into the sump pit and be pumped out or, if the topography around the house is sloped, the drainpipe can be pitched and run under the foundation to a low point on the exterior, which provides a free-flowing outlet.

High Water Table

I live in a rural area in Southern California that has a high water table and a lot of underground streams. With the recent heavy rains, one of the bedrooms has flooded due to water seeping up through the house's concrete slab. It has become a real mess. I realize that I am going to have to replace the carpeting and have considered a different type of covering such as wood parquet. Is there anything I can put on the concrete slab to seal it and prevent future occurrences?

Coating the bedroom's concrete slab to prevent water seepage will not provide a lasting solution to your problem. The hydrostatic pressure that develops after an extended heavy rain could lift or break the seal of paint or epoxy coating.

The best way to prevent the water seepage is to keep the level of the subsurface water below the underside of the home's concrete slab. You can do this by installing a sump pit in the slab with a pump that is automatically actuated by water level or water pressure. The depth of the pit should be sufficiently below the underside of the concrete slab to accommodate the pump and the associated piping and wiring. The capacity of the pump will depend on the rate at which the water table rises and may have to be determined by trial and error.

In order for a sump pump to work effectively, there should be a substantial gravel bed below the slab rather than compacted soil. With a gravel bed, water will flow readily into the sump pit when the pump is actuated, whereas with compacted soil, the water will not flow as readily and a sump pump may not be as effective. In this case, it would be necessary to install below the slab interconnected perforated drainpipes that discharge into the sump pit. Subsurface water entering the pipes would flow to the sump pit and be pumped out.

It is important that the pump discharge pipe terminate sufficiently far from the house so that the water will not accumulate around the foundation. Otherwise, the water will seep into the house.

I do not recommend covering the floor with wood parquet unless the concrete slab is 100 percent dry all year long. Even if you stop the seepage, there could be nonvisible dampness at the slab surface because of capillary action drawing up the moisture below. Dampness on the underside of the wood parquet will make the flooring cup and swell, causing it to warp.

SUMP PIT/PUMP

Sump-pump Care

Can you tell me how to care for my sump pump? Every winter I have to call a repairman after the first rain to get it started. What is the standard procedure to keep these things in good running condition?

Normally, a sump pump does not require any maintenance. Certainly not yearly maintenance. In all probability, the pump's intake port is sucking in dirt from the sump pit, and this is clogging the pump. The pit should have a liner, which prevents the earthen walls of the pit from eroding and depositing silt and tiny pebbles at the base. The pump should also be positioned so that the intake port is not at the base of the pit. Some sump pump manufacturers also make screens or filter boxes that can be installed in front of the pump's intake. Check with the manufacturer of your pump to see if such optional equipment is available. You can also make one yourself using a window screen. However, you should periodically check the screen to make sure it is not clogged.

Water-powered Backup Sump Pump

I just moved into a new house. The builder installed a sump pit and pump in the finished basement as a precautionary measure against water seepage. The last hurricane that passed through this area flooded a number of basements because power was knocked out and sump pumps were unable to operate. Is there anything that I can do to prevent a flooded basement in the event of a power failure?

You can install a water-powered sump pump in the same pit alongside the electric sump pump. The pump utilizes municipal water, which is available even during a local power failure. The pump's basic principle of operation is similar to that of an aspirator. That is, pressurized water flowing by a small opening will create suction. This, in turn, draws water out of the sump pit.

The cost of a backup pump system is expensive, but its cost is considerably less than the cost of repairing damage caused by flooding. The decision about whether

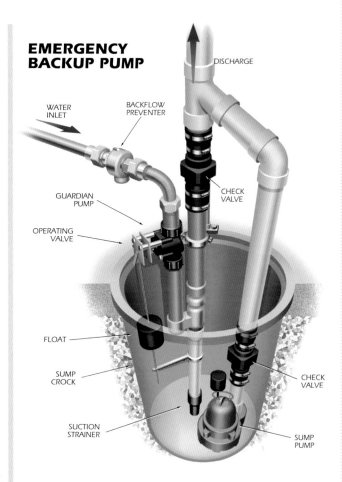

EMERGENCY BACKUP PUMP

DISCHARGE

WATER INLET

BACKFLOW PREVENTER

GUARDIAN PUMP

OPERATING VALVE

CHECK VALVE

FLOAT

SUMP CROCK

SUCTION STRAINER

CHECK VALVE

SUMP PUMP

to purchase a pump will depend on your tolerance of risk. It's like buying insurance. What is the probability that a power failure and flooding will occur simultaneously? If you live in an area where this happens often, it makes sense to consider a backup pump system.

Sump-pit Drainage

Is it normal for a sump pit to completely drain even though the pump suction is 6 inches above the bottom of the pit? The water is leaching into the ground and I do not know if this is normal or if I have pit problems.

It's not a question of normal or not normal. It is a function of the porosity and drainage properties of the soil under the house. The fact that the water seeps into the ground indicates that the drainage below the floor slab is quite good. In this case, if water accumulates under the floor slab it will percolate into the ground rather than build up in the sump pit and activate the pump.

Even though the soil drains well, a sump pump is helpful to remove water that rises up from under the slab due to melting snow or heavy rain.

Sump-pit Bottom

My basement has a sump pit with concrete sides and a sand bottom. In other words, the bottom is not lined with concrete or other material. When it rains, the water table rises, and my sump pump runs for four or five days straight. Would I have a problem if I closed the bottom of the sump pit?

Yes, you would have a problem. Your sump pit is working as it should. That is, it's designed so that the bottom allows the entry of subsurface water. If the bottom of the sump pit is sealed, then as the water table rises, the water will press against the underside of the basement floor slab and seep through the cracks and open joints that often exist in basements.

On the other hand, if there are no cracks or open joints, depending on the level of the water table, the pressure the water exerts on the slab (known as hydrostatic buildup) could cause the concrete basement floor slab to crack and heave.

With your system, by pumping out the water that builds up in the sump pit, you are controlling the level of the water table so that it doesn't rise to a point where it will press on the underside of the floor slab, thereby eliminating the problem of water seepage through the cracks into the basement. If the sump pit keeps the basement dry, our advice would be not to modify it.

Condensation on the Foundation

I have a home in the mountains of Arizona, and I find condensation collects on the beams bolted to the foundation. Vents have to be closed in winter to keep the pipes under the house from freezing. There is 2 inches of foam insulation on the inside of the foundation wall. What can I do to keep the water from accumulating on the 2 x 8 bolted to the foundation?

Moisture buildup in a crawlspace—assuming it is not the result of water seepage—is normally caused by subsurface water. Even when the ground in the crawlspace seems dry and dusty, moisture can accumulate in the area as a result of capillary action. Capillary rise occurs in nearly all crawlspaces built in areas where the soil is clay or silt. According to the University of Illinois Small Homes Council, as much as 18 gallons of water per day can evaporate into a crawlspace under a 1000-square-foot house.

If this is the source of moisture, then it can be controlled by covering the ground with a vapor barrier (4-mil polyethylene). The interior joints should be overlapped by a minimum of 6 inches, and the perimeter edges should be turned up onto the walls of the crawlspace.

Another possible source of moisture into a cold crawlspace is from the warm house above. Excessive moisture inside the house (high water vapor pressure) can travel downward through the floor into the cold crawlspace (low water vapor pressure) and condense on the cold surface. To prevent the downward flow of moisture vapor, you can install a vapor barrier between the overhead floor joists in the crawlspace.

Dehumidifying a Damp Basement

We have a moldy, musty odor in our basement that seems more pungent in the spring and fall. We have tried using mesh sacks filled with calcium chloride to reduce the odor, but they're ineffective. Occasionally, the sacks drip water. How can we decrease the odor?

The sacks hanging in your basement absorb moisture that can lead to mold and mildew growth, but they are not as effective as a mechanical dehumidifier. You need an electric dehumidifier. These appliances have a humidity control and will run continuously until the humidity setting is reached. The water that they condense out of the air, known as condensate, collects in a tank. You have to manually empty the tank, pump the condensate to a disposal drain, or drain it with a hose to a floor drain.

Moisture problems such as yours are more apparent at times of high humidity. During warm weather, the basement is cooler than the outdoors. Warm air can hold more water vapor than cool air. When warm outside air works its way into the basement, its temperature drops and causes the relative humidity in the basement to increase. Aside from contributing to mold and mildew growth, this makes the basement air feel clammy and uncomfortable during the summer.

Humid Basement

Our basement is quite humid most of the time, although water does not seep into it. Is it helpful to coat the foundation wall using a cement based paint, or will a good quality water based paint work as well?

It's not unusual for a basement to become humid, especially during the warmer months. A basement, by its very nature of being below grade, will be cool. Cool air cannot hold as much moisture as warm air, and, as a result, the humidity level increases when the warm outside air works its way into the basement.

I doubt that the humidity problem in your basement stems from water vapor permeating the foundation wall. You can check this by taping a 6- × 6-inch piece of aluminum foil to the wall. Leave it there for 24 hours and then remove it. If the side of the aluminum facing the wall is damp or has beads of condensation on it, then your assumption is correct.

However, coating the foundation walls with a cement based paint will not solve the problem. Cement based paint is formulated to stop water seepage through the foundation, but because it "breathes" it will not stop water vapor. Water based paint will not stop water vapor either, and an oil based paint will form a film on the wall that will eventually blister and peel.

To prevent water vapor from coming through the foundation wall, you will have to cover the foundation on the exterior side with nonpermeable membrane, or with a tar or asphalt coating. Another option is to remove the water vapor from the air by using one or more dehumidifiers. However, you must let the appliance run until its hygrometer setting shuts it off. Over the years, I've had clients call me and say that their dehumidifier was not doing its job even though they let it run for several hours each day. Depending on the capacity of the dehumidifier and the size of the basement, the dehumidifier may need to run continuously for many hours.

Hot TV Room

I have a TV room in my basement. The room is hot during the summer months. We have a dehumidifier in the room, so we thought the room should be cool. Could you please tell me why it is hot down there?

The room is hot because of the dehumidifier. A dehumidifier is basically a small self-contained air conditioner. An air conditioner discharges the heat removed from the circulating air and from its compressor to the outside, but a dehumidifier dumps that heat into the room.

If the TV room is small and the dehumidifier runs continuously, it discharges enough warm air to heat the room. A dehumidifier is not used to cool a room. It makes a room more comfortable by lowering the relative humidity.

You would be better off with a small wall-mounted air conditioner. This unit also removes excess humidity and cools the room. Even though the TV room is in the basement, a section of the foundation wall is generally above grade. If it is a concrete block wall, an opening can be cut in the foundation for an air-conditioning sleeve, or an air conditioner can be installed in a basement window.

Damp Basement Closet

I hope you can help me with a very aggravating problem. About a year ago, I built a closet in the basement to store our clothes in the off-season. Our summer clothes were stored last winter and they were fine this spring when I took them out. However, when I retrieved our winter clothes, which had been stored for the summer, I found them full of mold and mildew. How do I solve the problem?

Basements, because they are below grade, are cooler in the summer than the rest of the house. As a result, the moisture in the humid summer air tends to condense in the basement, making that area quite damp. This, in turn, promotes the growth of mold and mildew. Opening the windows and using a fan to circulate the air will only work when the humidity of the outside air is not as high as it usually is in the summer.

And, the problem is compounded in a closet because of the confined space and stagnant air.

The best approach is to install a dehumidifier in the basement. While it's not practical to place it in the closet, you can install vent openings in the top and bottom of the closet door to help circulate the basement air.

Another approach is to use chemicals that absorb moisture such as silica gel and activated alumina. These have the capacity to absorb half their weight in water. They can be placed in the closet in a bucket or cloth bags hung from the closet pole. After they've become saturated, the water can be drawn off by heating and the chemicals can be reused.

Damp Basement Room

I am interested in making a below-grade finished basement into a comfortable office. There is no water seepage, but even so, the basement is cold and clammy in the winter and humid and musty in the summer, I'm using a dehumidifier and a portable heater, but cannot overcome the discomfort. Any suggestions?

You have the ingredients for making the finished basement into a comfortable office all year long—a dehumidifier for the summer and a space heater for the winter. However, either you need additional units or you are not using the units that you have for a sufficient time period.

If you have two or more rooms in your basement, a single dehumidifier in one room will not effectively reduce the humidity in the other room, especially if the door between the rooms is closed. If the door is open and you want to use a single unit you will need a fan to provide sufficient air circulation between the rooms.

Most humidifiers are equipped with an automatic control switch, which will shut off the unit when a predetermined humidity setting is reached. In order for the humidifier to wring out a sufficient amount of moisture from the air to reach a comfortable level, you must let the unit operate until it shuts itself off. During hot, muggy summer days the dehumidifier may run almost continuously. If you allow the dehumidifier to run for only one or two hours a day, it will not reduce the relative humidity to the comfort level.

A space heater is effective in heating a single room. If you want to heat more than one room, you will need a fan to circulate the air between the rooms. Even in a single room, a fan will be helpful in achieving a uniform heat distribution. A thermostat that is mounted on a wall across from the heater, rather than on the heater itself, will also improve heat distribution within the room.

Mold Creeps In

I have just noticed this since I had my rooms remodeled. I had them paneled, and I had insulation put on first. Now I find mold spots forming inside glass picture frames on the shelves. My basement is a dirt floor under the living room. Could the dirt floor be causing this problem? Any help would be appreciated.

I believe this problem is caused by the dirt floor in your basement. Even when the dirt feels dry to the touch, it wicks up subsurface water, and this is released into the area under the living room and eventually into the living room itself.

Before you remodeled your rooms, there apparently were enough open joints in the walls through which the moisture could escape to the outside. After you remodeled, those joints were sealed, causing the moisture to remain in the rooms.

In order to control the moisture buildup, you should cover the dirt floor in the basement with a vapor barrier, such as 4- or 6-mil-thick polyethylene plastic sheets. Overlap the sheets and tape the joints shut.

Persistent Musty Odor

What can we do to rid our home of the persistent musty odor in the basement? Books and other items stored there develop a smell that is usually retained. We have tried fans and dehumidifiers, to no avail.

A musty odor is quite common and is caused by mildew, a tiny, simple plant also known as fungus or mold. Mildew grows wherever it is damp, dark, and poorly aired. It also feeds on cotton, linen, wood, and paper.

Mildew can be prevented by keeping an area or an item dry, usually with adequate air circulation. In a basement, this is often done by decreasing humidity with one or more dehumidifiers or by heating the basement. Mildew can be removed from an item using chlorine bleach, but clean a test patch first to determine whether the bleach will damage the item.

Mold spores can appear as black, brown, blue, orange or white specks, but they are not always visible to the naked eye. They often grow in carpeting, upholstered furniture and even on the back of wall paneling. If you've done all of the obvious things to eliminate the musty odor, and it persists, it's because mold is growing in areas that are not readily visible.

At this point, you may have to call a company that specializes in treating mold conditions in "sick houses," a phrase that describes houses with a range of air-quality problems. Unfortunately, these companies are scarce.

Beating Below-Grade Mildew

I have a second home, which we use mostly on weekends. The house is four years old and is built into the side of a hill. My problem is mildew in the closet, bathroom and laundry room on the lower floor. Last summer everything in the closet mildewed—clothes, shoes, walls. The bathroom and laundry also mildewed around the baseboards, and on some of the walls.

We washed the walls and ceilings with a bleach solution, which seems to retard growth of mildew but doesn't stop it. I've been told to vent the area by putting in small fans, or to run a dehumidifier. But since it's only a weekend house, I don't want to leave anything electrical on when the house is empty. I have found this to be a common problem with weekenders all over our area, but no one seems to know what to do.

Mildew thrives in a damp environment, and in order to prevent it, it's necessary to control the dampness. When the dampness is the result of condensation of the warm, moist summer air, it can be controlled by a dehumidifier and ventilation as was suggested to you.

However, it sounds as though your problem is caused mainly by moisture buildup on the foundation walls and floor slab because of the hydraulic pressure of wet soil adjacent to the house. In this case, additional measures must be taken.

Since your house is built into the side of a hill it's likely that the ground on the uphill side is not graded properly. The ground, for at least several feet, should slope away from the house so surface water will not accumulate against the foundation. Also, gutters and drain pipes should channel roof rain runoff away from the house,

and toward the downhill slope of the land, rather than letting it saturate the soil adjacent to the foundation.

Since you don't want to run a dehumidifier and fan during the week when the house is empty, you might try getting rid of the dampness using chemicals that absorb moisture, such as silica gel and activated alumina. These chemicals can be placed in open containers or cloth bags in the problem areas. The chemicals have the capacity to absorb half their weight in water. After they have become saturated they can be heated to draw off the water, and then reused.

Overflowing Window Well

Please advise us on the problem of an overflowing window well. When there is heavy rain, the water builds up in the well, goes through the window frame and flows into the basement. How much crushed rock should be placed so the water will seep through?

You should prevent the rain from entering the window well rather than let it seep through a gravel bed. You can buy clear plastic domes that fit over the window-well opening at home centers.

If your window well is an odd size, you can build the equivalent of a plastic dome. Buy a sheet of clear plastic, and cut it to size so that it overlaps the sides of the window well by about 6 inches. Install a ledger board on the wall above the window well, and lay the plastic so it is inclined with the top resting on—and secured to—the ledger. The bottom of the plastic should overlap the outside edge of the window well by a few inches.

If the window well is filled with gravel, during a sustained rain, water will seep down through the gravel and accumulate around the base of the foundation, where it could seep into the basement via a crack in the foundation wall.

Floor Tiles Come Loose

Most of the resilient floor tiles on my basement floor are lifting at the corners. Underneath the tiles is a saltlike substance. The tiles were placed just five years ago, and this condition is occurring on about one-third of the floor. Is there a solution?

In most cases, this is caused by moisture rising up through the floor. The white substance is indeed a salt. It's called efflorescence. As moisture passes through masonry, it dissolves salts present in it. When the moisture evaporates, salt crystals are deposited on its surface.

The condition is probably caused by water accumulation under the basement floor slab. It may help to identify and eliminate the causes of the accumulation. Downspouts that empty roof runoff too close to the foundation, improper grading around the house and clogged foundation footing drains can all cause water to build up under the floor.

It's hard to say whether the basement floor will be suitable for tile after the moisture problem is corrected. Residual moisture that remains may cause tiles to lift and may make it difficult to establish a bond firm enough to hold the tile in place.

If you want to tile the floor again, the loosened tiles will need to be removed and the surface below cleaned and prepared to ensure a strong bond. Consult with a flooring store about which products to use.

FOUNDATION

Loose Sills

I live in California in a house built in 1926. The mudsills are not bolted to the poured concrete foundation. If what I read about earthquakes is true, I'd better get them fastened down. I've been told that there's no simple way to do this. How can I secure them without jacking up the whole house to get the bolts in?

For those not familiar with the term, the mudsill, or sill plate, is the lowest member of an exterior wood-frame wall, which rests on the foundation and supports the joists and upright portions of the frame. The term mudsill originated from the procedure of correcting irregularities in the top of the masonry foundation by embedding the sill in a layer of grout or fresh mortar. Normally, this sill is anchored to the foundation wall.

I agree that you should fasten the mudsill to the foundation. In the event of an earthquake, the house could slide, shift or even overturn. Because of the limited space, installing anchor bolts in the top of the foundation would be difficult and costly. Jacking up the house is not a practical solution and not recommended. I suggest that you have an ironworks shop fabricate iron angle brackets. The brackets should have a small spike on the short end which can be hammered into the sill plate to prevent sliding. By mounting brackets on all of your house exterior walls, you will adequately secure the structure. Specifications as to the bracket size and spacing will depend on your locale. So, check with a licensed professional engineer in your area specializing in structural design.

SEAM IN FOUNDATION

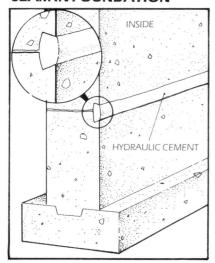
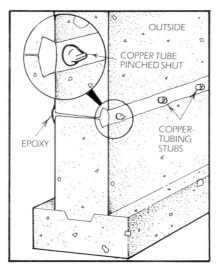
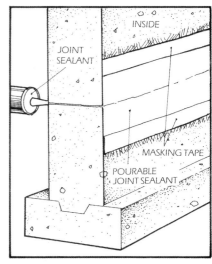

Serious Seepage

When it rains, I have a leak in my foundation wall along a cold joint that runs around the foundation's perimeter. The joint is the result of pouring the concrete on two different days. To fix this and make the wall more attractive, I thought of troweling a portland cement/sand mixture over the wall, but will it stop the seepage?

A cement stucco can improve the wall's looks, but won't stop the leak. The surest (and most expensive) fix is to intercept the water before it gets to the wall with drainage, then undercut the crack with a masonry saw or chisel, and seal the seam inside and out. A less expensive fix is to patch only the seam's inside face.

There are at least three choices of patch material: hydraulic cement, epoxy patch and flexible joint sealant. You can buy hydraulic cement in hardware stores. However, you will probably have to order the epoxy patch. Joint sealant may be sold at home centers or it can be ordered.

Hydraulic cement is simply troweled into the widened crack. Epoxy patch is troweled on both sides of the wall, though the seam is widened only on the wall's outside. If the leak is serious, you could patch the seam, as shown, and install copper-tubing stubs that form injection ports. Thin epoxy is injected through the tubes, which are later pinched shut, cut off and covered.

Joint sealant is squeezed into the enlarged seam after the seam is wire brushed. Apply butyl rubber, silicone or (preferably) polyurethane joint sealant using a caulk gun. If possible, on the inside, mask off along the seam and brush on a pourable (also known as self-leveling) joint sealant between the pieces of tape. You can temporarily repair the joint by just wire-brushing its inside face and brushing on joint sealant.

Floor Slab

The concrete floor of our basement is constantly covered with a powderylike dust and nothing that we have tried prevents it, including applying a concrete sealer several years ago. Vacuuming and mopping helps for a week only. Any ideas what this would be and how can we get rid of it?

The condition that you describe is called dusting. According to the Portland Cement Association, the dust consists of fine particles of concrete aggregate (sand and stone). The particles form a thin and loosely bonded layer on the concrete's surface. Movement across the concrete breaks the particles loose and stirs them into the air.

Several things can cause this, such as using a concrete mix with a low cement content or one that contains too much water. It may also result from improper finishing and installation, such as subjecting the concrete to freezing temperatures before it has cured, allowing the concrete to dry too rapidly or troweling it smooth while it is still so wet that water is standing on the surface.

One way to correct the dusting is to grind off the thin layer to expose the solid concrete underneath. This, however, may not be very practical in your home. Another method is to apply a chemical surface hardener that contains either sodium silicate, commonly called water glass, or a metallic fluosilicate (such as magnesium and zinc fluosilicate). These products are available at construction supply companies.

CRAWLSPACE

Drying Out a Damp Crawlspace

I live in a two-story home built over a 3-foot-high dirt-floor crawlspace covered with heavy plastic sheeting. The space is vented on all four sides, with a total of 17 vents. The problem is that when the outside temperature is about 90°F and the humidity is high, moisture condenses on the insulation between the floor joists. Puddles form on the plastic sheeting and mildew grows on the exposed portions of the joists. How can this be remedied?

According to accepted design practice, your crawlspace is well ventilated. In fact, it has more vent openings than is required. Condensation in a crawlspace

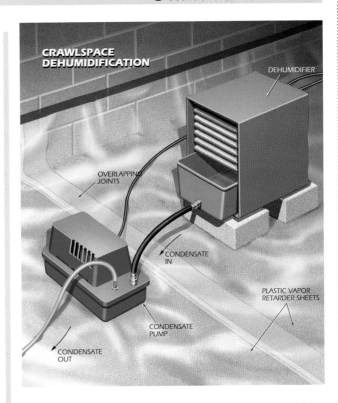

CRAWLSPACE DEHUMIDIFICATION

DEHUMIDIFIER

OVERLAPPING JOINTS

CONDENSATE IN

PLASTIC VAPOR RETARDER SHEETS

CONDENSATE PUMP

CONDENSATE OUT

during the summer is a very common problem. A crawlspace that is partially set in the ground will always be at a lower temperature than the outside air during the summer. It's normal that the moisture in the warm outside air will condense when it enters this space.

However, quite often condensation in a crawlspace is not caused by warm humid outdoor air, but is the result of inadequate insulation on cold-water pipes and air-conditioning ducts. Also, the insulation should be covered by a vapor barrier, but frequently it isn't. Assuming that your problem is that of warm, humid air entering the crawlspace, you could reduce it by closing the vents whenever the temperature and humidity are high enough to cause condensation to form, but this is an impractical solution.

A more practical approach is to install a dehumidifier. These appliances are controlled by a humidistat. Once you set the machine, let it run until it shuts itself off. The condensate that accumulates at the bottom of the pan can be directed into a sump pit, if there is one, or to a small condensate pump where it can be pumped to the outside. Dehumidifiers are available at home centers and department stores. You can purchase condensate pumps through industrial supply houses and plumbing supply houses.

Another method of controlling the humidity in the crawlspace is to install an exhaust fan operated by a humidistat.

Sweaty Ducts

The air-conditioning ducts in the crawlspace under my four-year-old home sweat so badly during the summer that small puddles form on the ground below them. How can this be prevented?

First, take steps to reduce the moisture buildup within the crawlspace. If there is an exposed-earth floor, cover it with a vapor barrier such as 4- or 6-mil polyethylene sheets with edges overlapped and taped together. Tape the sheets to the foundation wall. Look for leaks in plumbing pipes, and repair them. Also, make sure that the gutters are clean so that water will not overflow and seep into the crawlspace. Finally, check to see if the dryer vent discharges moist air into the crawlspace.

After taking steps to reduce moisture vapor, you must also insulate the ducts and seal the insulation with a vapor barrier. The surface temperature of the ducts is below the dew point of the air in the crawlspace. Insulating the ducts without covering the insulation with a vapor barrier will not prevent condensation from forming on the ducts. By covering the insulation with a vapor barrier, such as polyethylene sheets with sealed joints, you can prevent the moisture in the air from coming in contact with the cool surface of the ducts, and then condensing.

Duct Leakage

The metal ductwork for our heating/cooling system is in our crawlspace. It has no large gaps, but there is some leakage where the pieces are joined. Duct tape was used sparingly. Should we cover all the open joints with duct tape? Also, should the ducts be insulated?

Open joints and inadequate insulation in ducts are major sources of wasted energy, especially when the ducts are in a crawlspace. An easy way to check for nonobvious leakage is to wet the palm of your hand and hold it over the suspected area. If there is a leak, you will feel a cool sensation on your palm. All leaks should be sealed with duct tape, then double-checked for leakage.

The ducts are probably insulated. Metal ducts are sometimes insulated on the inside. However, duct insulation is usually marginal, and additional insulation is needed to reduce heat loss in the winter and to keep condensation from forming when the air-conditioning system is running. When insulating metal duct from the outside, it's necessary to wrap a vapor barrier around the insulation to prevent condensation from wetting the insulation.

Crawlspace Concern

My house is new, and built over a gravel covered crawlspace. The builder says that a plastic sheet vapor barrier is not necessary, that the gravel itself is a sufficient barrier. What is your opinion on this?

I strongly disagree with your builder. There is the possibility that in the future there will be a problem with the roof gutters or downspouts (to cite one example) that may cause water to flow into the crawlspace.

Even though the ground under the gravel seems dry, moisture can travel upward from lower layers of soil by capillary action and evaporate into the crawlspace. According to the University of Illinois Small Homes Council, this can bring 18 gallons per day into the crawlspace under a 1000-square-foot house.

I suggest you use polyethylene sheets with all joints sealed by plastic tape.

Buckled Floor

We have a moisture problem. Our four-year-old, two-story home is a few hundred feet away from a saltwater creek. By midsummer, condensation develops in the crawlspace under the house and drips down on the plastic vapor barrier that covers the crawlspace floor. By autumn, the dampness has caused the oak flooring in the room above it to buckle. When the heat comes on in the winter, the floor dries out and settles down, though not completely. Can you help us solve this problem?

Check that the vapor barrier on the crawlspace floor has no holes or open joints. Specifically, check the joints between the vapor barrier and the foundation walls. All open sections must be sealed with duct tape.

In addition, in the winter when humidity is low, install a vapor barrier (large 4-mil polyethylene sheets with taped overlapping joints) to the underside of the floor joists in the area below the room with oak flooring. Also, place a dehumidifier in the basement that can discharge condensation to the outside by means of a condensate lift pump. Disconnect the dehumidifier during the winter.

It's important to keep the crawlspace dry, not only for the oak flooring, but because excessive dampness promotes rot and creates conditions conducive to termite infestation.

Vapor Barrier Advice

I have received conflicting advice regarding vapor barriers for crawlspaces. Some advise leaving small spaces between the sheets to allow the ground to dry out under the barrier. Others advise leaving no gaps. Also, should the barrier extend up the concrete walls?

In a crawlspace, the vapor barrier is usually made up of 4- to 6-mil-thick polyethylene sheets with overlapped joints that are sealed with heavy-duty plastic tape. The sheets are run several inches up the sides of the foundation and are taped to the wall.

You should not leave spaces between pieces of the vapor barrier. To be effective, it must be continuous. The spaces would allow moisture vapor into the crawlspace. Some background is helpful in understanding how a vapor barrier works. It is installed to stop the capillary rise of moisture in the soil from becoming airborne vapor. It can't prevent water from collecting in the soil under the crawlspace.

If you find that the ground in the crawlspace is wet, you should take measures to dry it out. For example, the ground should slope away from the house on all sides, drains should be installed to move water away from the foundation and downspouts should discharge water far enough from the house that it can't seep into the crawlspace.

I have a house above a crawlspace. The space has a plastic sheet installed over it. I would like to insulate the house's floor. Should I use insulation with or without a vapor barrier? If it needs a vapor barrier, should it be on the top or bottom of the insulation?

First, you should use insulation that has a vapor barrier on one side. When you install the insulation, the vapor barrier should be on the top so that it faces the underside of the floor above, rather than facing the crawlspace floor. Water vapor movement occurs because of a difference in vapor pressure. During the winter months, the vapor pressure in the heated rooms above the crawlspace is normally higher than in the crawlspace. Consequently, to prevent moisture movement into the crawlspace, and condensation, the vapor barrier is placed facing the heated portion of the structure.

During the summer months, in many areas of the country, the vapor pressure in the crawlspace is greater than in the rooms above, especially if those rooms are air-conditioned. In some cases this can cause problems because the vapor will move up toward the rooms. Depending on the temperature and corresponding dew point, the vapor might condense when it comes in contact with the vapor barrier above the insulation. In this case, the insulation can become saturated, sag and eventually fall.

To reduce the chances of moisture vapor producing wet insulation, you should check that there are no gaps between the vapor barrier sheets that are installed over the floor of the crawlspace. The sheets should be taped together, and they should be taped to the foundation wall. Also, the space should have adequate vent openings so that there is cross ventilation. If the vapor problem is severe, install a dehumidifier.

Crawlspace Moisture

The sandy soil of the crawlspace below my house is usually moist, and downright wet during the rainy season. When a termite exterminator was treating my house, he placed a plastic sheet over most, but not all, of the space. Is this adequate?

The sandy floor of the crawlspace should be completely covered with a vapor barrier comprised of a polyethylene plastic sheet 4 to 6 mils thick. This reduces moisture in the space, as well as the potential for rot in the wood framing. Dry conditions also reduce termite and carpenter ant activity.

The vapor barrier must cover the entire floor, and joints between sheets should be taped closed with heavy-duty plastic tape. The vapor barrier should extend to the foundation wall and be taped to it.

You can reduce crawlspace moisture by making sure that gutters and downspouts discharge water away from the house, and that the ground slopes away from the house on all sides. Furthermore, there should be at least two vents—one each at opposite sides of the foundation. The vents should be sized as follows: 1 square foot of unobstructed vent area for each 1500 square feet of crawlspace floor.

A damp crawlspace should be inspected periodically. Make sure that the vapor barrier is intact at all points. Also, you should check for insect activity. If you notice standing water under the sheets, consider installing a sump pump and a foundation drainage system. If your house already has a drainage system, perhaps it needs repairs.

Dripping Floor Joists

I live in a three-bedroom ranch home built over a crawlspace. Two years ago, I put in air conditioning. Yesterday, I discovered there was so much moisture in the crawlspace that it was dripping from the floor joists. What can I do about it?

It is possible that the problem is unrelated to the installation of the air conditioning. Check the water distribution piping and drain lines in the crawlspace for leaks. Also, the vents around the perimeter should be clear. Quite often the vents' insect screens get clogged with dust and dirt.

If you have a dirt floor in the crawlspace, cover it

with a plastic sheet. Any air-conditioning ducts in the crawlspace should be insulated. And any open joints, through which cold air can escape and condense, should be sealed with duct tape.

Crawlspace Ventilation

As an architectural designer/draftsman, I've had problems convincing homeowners and contractors of the importance of proper ventilation and vapor barriers. However, I was concerned about the misprint of 1 square foot of unobstructed vent area for each 1500 sq. ft. of crawlspace floor as mentioned in a previous "Clinic." Current Uniform Building Code (UBC) and state specifications require 1 square foot per 150 square feet of crawlspace floor area.

There are two conditions that govern the amount of ventilation required for a crawlspace. If the floor in the crawlspace is exposed to earth, then the net-free vent opening must not be less than 1 square foot for each 150 square feet of crawlspace floor area. However, when the floor is covered with an impervious membrane, such as an approved vapor barrier, like 4- or 6-mil polyethylene sheets, then the net-free vent opening may be reduced to 10 percent of that figure, or 1 square foot of vent opening per 1500 square feet of crawlspace floor area. My recommendation in the column that you cite included the use of a vapor barrier.

Crawlspace Water

The crawlspace under my six-year-old home is covered with pea gravel that has a plastic vapor barrier on top of it. Throughout most of the year, the pea gravel is damp. At other times, small areas of gravel are covered by at least an inch of water on top of the vapor barrier. A certified home inspector has assured me that this is not unusual for a crawlspace and that I need do nothing about it because there are no water marks on the foundation walls and the floor framing shows no signs of decay. However, several waterproofing firms insist that I need to dry up my crawl space by installing a perimeter drain and a sump-pump system. The cost ranges from $2500 to $8000. What would you advise?

A crawlspace should be kept relatively dry. Why? Because continuously wet conditions are conducive to decay fungi, which will cause rot. The dampness also promotes mold and mildew and is very inviting to carpenter ants and termites.

Certainly, installing a sump pump with perimeter drains will control the water condition. However, make sure that the plastic vapor barrier is properly installed before going through the expense of having a drainage system put in. The plastic ought to completely cover the gravel, even along the perimeter of the foundation. Its edges need to be sealed to the foundation and its joints should be taped. If all this is done, the vapor barrier should prevent excessive moisture in the crawlspace.

In those areas where water accumulates on top of the vapor barrier, check to see if the water is seeping through open or loose vapor-barrier joints or through the foundation wall. If water is coming through the walls, check the grade of the ground adjacent to the house, which should slope away from all sides of the foundation. Also, monitor the effectiveness of gutters and downspouts. Furthermore, if the ponded water is the result of condensation dripping from the overhead cold-water pipes, you can control this problem with a dehumidifier.

If you can't maintain a tight seal in all the vapor-barrier joints and control water seepage through the foundation, you will need to install perimeter drains and a sump pump.

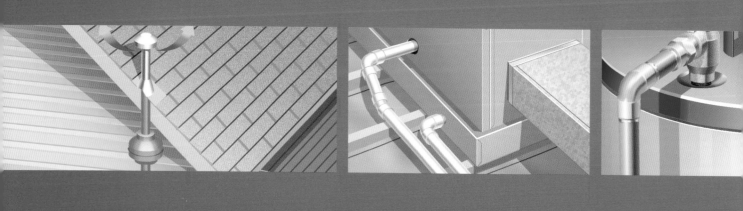

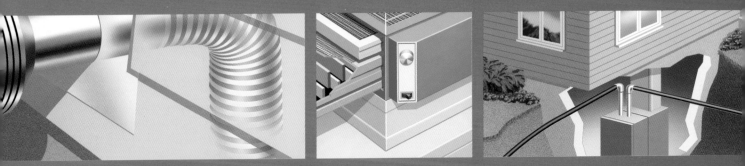

Electromechanicals

11. Electrical System

ELECTRICAL UPGRADE • GROUNDING • GROUND-FAULT INTERRUPTER • ALUMINUM WIRING • LIGHT FIXTURES, SWITCHES • DOORBELL • WIRING

ELECTRICAL UPGRADE

Electrical-Service Upgrade

Our 25-year-old double-wide ranch home has a 100-amp electrical service box with circuit breakers. We'd like to upgrade the service box to 200 amps. Is this possible? Is there a lot of work involved?

Homeowners who have embarked on a major remodeling job—especially those involving the kitchen—quickly learn that older homes are electrically underpowered by modern standards. Many have only 60 or 100 amps of electrical service, while most newer homes have 200-amp service. Older houses generally did not have a trash disposal, dishwasher, hot-water dispenser or multiple countertop outlets, all of which are common in modern kitchens.

To determine the extent of the electrical work necessary, you should hire a licensed electrical contractor. The contractor will examine the house's electrical system and perform load calculations. In some cases, the contractor may determine that the existing panel box can accommodate the new kitchen. But in other cases, a new subpanel box may be necessary. Large, complex remodeling jobs may require the installation of a new higher-amperage service cable and a larger higher-amperage panel box.

Specifically, if the inlet electrical service to your house is 100 amps, you cannot upgrade to 200 amps merely by changing the panel box to one rated for 200 amps. You must have the utility company bring a 200-amp cable to the weatherhead connection on the house. Next, your electrician must run a 200-amp cable from the weatherhead to the electrical meter, and from the meter to the new panel box. The branch circuits from the old box are relocated to the new box and additional branch circuits are added.

All electrical work must meet local, state and national electric codes. When the job is done, the electrical contractor will provide you with a certificate of approval from the municipal inspection agency.

Upgrading Receptacles

I have an older home that was wired before the days of 3-wire grounded circuits. Would it be safe to replace the old receptacles with new, 3-slot receptacles and simply use a short wire to connect the grounding screw to the neutral wire?

Your proposed solution is unsafe and can result in severe electrical shock. It definitely should not be used. Although the neutral line is grounded at the panel box, it is not a grounding wire and, in fact, is a current-carrying wire under normal load conditions. By connecting the grounding screw on a receptacle to the neutral wire, you are making the appliance chassis electrically hot. The potential exists for an electric shock that could be fatal.

Having 2-slot receptacles doesn't mean that your receptacles are not grounded. Unless knob-and-tube wiring or old 2-wire Romex was used, your receptacles are probably grounded. You can check it out with an inexpensive neon circuit tester available at home centers and electrical supply stores.

Put one leg of the tester on the receptacle faceplate screw. If necessary, scrape off any paint to make good contact. Put the other leg in the hot slot.

If the receptacle is grounded, the neon bulb will light. In this case, a simple 3-slot adapter will do the job. Plug in the adapter and connect the pigtail to the faceplate screw.

If your system is not grounded, then I recommend rewiring those receptacles that are used for appliances.

GROUNDING

Electrical Grounding

In our previous home, our electrical system was nicely grounded to the copper water pipe where it came through the foundation. We have well water in this house, and I just noticed that the electrical system's ground wire is clamped to the brass pipe coming from the well's pressure tank. What concerns me is that the pipe is immediately connected to a plastic waterline coming from the well. Would it be better to drive a copper rod into the ground and connect the electrical system's grounding wire to it? How deep must the ground rod be driven?

If the electrical system in your house is not grounded by any other means than the one you describe, then your electrical system is potentially hazardous. A grounding system provides protection against electric shock by directing current from a short to ground rather than through a person.

To correct the problem, you can use a $^5/_8$-inch-

diameter copper or steel ground rod. The rod should be at least 8 feet long and driven into the ground so its tip is at least 8 feet below the surface. Or, it can be buried in a trench that is at least $2^1/_2$ feet deep.

Depending on the electrical resistance of the ground, you may need more than one ground rod. Also, a special clamp that is rated for outdoor use is required to attach the ground wire to the rod. The wire's size (its gauge) depends on the distance from the ground rod to the main panel box. Typically, 6-gauge copper wire is used. In addition, the ground wire must be installed in a continuous length without any splices or joints.

There are numerous specific details about grounding

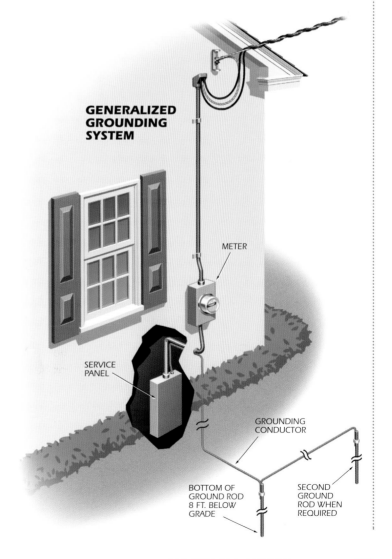

GENERALIZED GROUNDING SYSTEM

METER

SERVICE PANEL

GROUNDING CONDUCTOR

BOTTOM OF GROUND ROD 8 FT. BELOW GRADE

SECOND GROUND ROD WHEN REQUIRED

in the National Electrical Code. If you are not familiar with them, you should have the work done by a licensed electrician or, at the very least, have your municipal electrical inspector check your work. Check with your local building department for information.

Finally, it's important to know that all metal components in your house must be grounded, especially a metal plumbing system. Wherever there is the possibility for a break in the continuous length of a metal water pipe, the two ends should be connected with a bonding jumper wire. As an example, the water heater's cold-water inlet and hot-water outlet pipes should be connected with a bonding jumper wire. Otherwise, during the period when the water heater is being removed and replaced, the plumbing system will not be grounded.

Electrical Ground Versus Neutral

I have a question about the ground wire in a house's electrical service. At the service panel all black wires are connected to one side of the incoming service. All the neutral and ground wires are connected to the other side. Why is a separate ground wire required, as both the neutral and ground are connected to the same bus?

The purpose of a separate grounding wire in a circuit is safety. This wire does not carry current during normal operation. Instead, it connects to ground components such as metal outlet boxes in which receptacles and switches are installed, metal appliance cabinets, and frames of motors. Although the ground wire does not normally carry a current, it will carry current in case an electrical short develops. In this case, the grounding wire directs the current safely to ground rather than shocking the person touching the appliance.

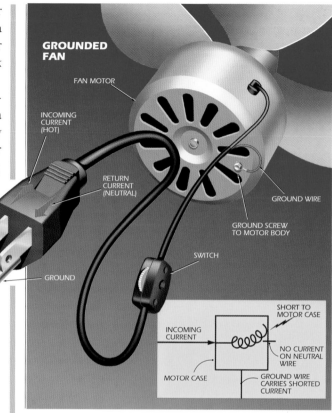

GROUNDED FAN

FAN MOTOR

INCOMING CURRENT (HOT)

RETURN CURRENT (NEUTRAL)

GROUND

GROUND WIRE

GROUND SCREW TO MOTOR BODY

SWITCH

INCOMING CURRENT

SHORT TO MOTOR CASE

NO CURRENT ON NEUTRAL WIRE

MOTOR CASE

GROUND WIRE CARRIES SHORTED CURRENT

The neutral wire, although grounded, is not needed for safety. It carries current and is needed to close the electrical circuit. This allows the current in a branch circuit to flow from the electrical service panel through the appliance and back to the grounding bar in the service panel.

Going to Ground

I recently purchased a personal computer, and need information on the proper way to protect my investment. I wish to use a surge protection device that requires a grounded three-prong receptacle. I presently live in an older building that doesn't have the grounded outlets. How can I safely connect my equipment? Because I will not be living at the

above residence for long, I need a simple and inexpensive solution. So far, no electronics store has been able to solve my problem. My only thought was to run a ground wire from the nearest cold water pipe to the outlet I intend to use.

If BX cable was used in wiring your building, then the outlet boxes in which the receptacles are mounted are grounded and solving your problem is easy.

You can check to see if the box is grounded with a neon circuit tester available at electrical supply stores. Stick one end of the tester in one of the slots of the receptacle and hold the other end on the cover plate screw. If the screw is painted over, scrape it clean for good contact. The neon tester will light if the box is grounded. If the first slot you try doesn't light the tester, try the other one. Only one slot is hot.

If your box is grounded, you can solve your problem with an adapter that converts a 2-slot outlet to three slots. It has a ground connector that must be fastened under the cover plate screw. If the outlet box is not grounded, you can still use the adapter. Fasten a wire to the ground connector on it and clamp the other end of the wire to the nearest water pipe.

Then, check the adapter (as described above) to make sure it's grounded. If it isn't, there's discontinuity in the water pipe ground. You'll have to locate this point and connect a jumper wire across it.

Lightning Zap

In one year's time, I've had two television sets damaged by lightning. It blows the tuner out even when the set is turned off, and the power cord is plugged in. A 15-amp voltage surge protector installed between the wall plug and the TV plug hasn't helped. We have cable TV that is grounded outside. What can I do to prevent this from happening again?

Your problem sounds like the classic example of what happens when the cable TV coaxial cable is not grounded properly.

The National Electrical Code specifies that the CATV cable should not have a separate ground, but that it should share a common ground with the inlet electrical service to the house.

If the electrical system and cable TV are grounded separately, a lightning surge can generate a difference of many thousands of volts between the power cord and the coaxial cable, causing the TV set to blow out. You can avoid this problem in the future by making sure the ground for the TV cable is connected and properly bonded to the main ground for the electrical service to your home.

GROUND-FAULT INTERRUPTER

GFI Protection

The electrical system in my house is controlled by circuit breakers. I was recently told to install a GFI circuit for my bathroom. Will this give me any more protection than the circuit breaker that already controls the bathroom outlet?

Yes, it will because they serve different functions. A circuit breaker or fuse will prevent a fire that's a result of excessively hot wires. This is done by automatically interrupting the circuit when the amperage it's carrying exceeds the capacity of the circuit.

On the other hand, a GFI (ground-fault interrupter) is designed to prevent a fatal electric shock by -

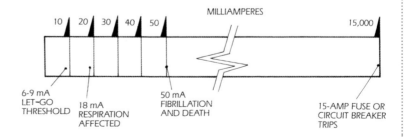

interrupting the circuit whenever there's an imbalance in the current between "hot" and "neutral" lines. The GFI will open the circuit when the imbalance is as small as 5 milliamps. This is 1/300 of the current needed to trip a breaker set for 15 amps. A GFI circuit interrupts the power within 1/40 of a second or less—fast enough to prevent injury to anyone in normal health. At 50 milliamps (1/40 of an amp), it takes only 3½ seconds for a person's pulse to stop.

Electrical Safety

Occasionally, while inspecting homes for prospective buyers, I find GFCI (ground fault circuit interrupter) outlets that are installed incorrectly, posing a risk to homeowners. A GFCI outlet is intended to provide ground fault protection to devices that plug into it. An added benefit of a GFCI is that it can extend that protection to outlets that follow it on the same circuit.

If it is wired incorrectly, however, the GFCI outlet itself will not have ground fault protection, but those that follow it will.

Some background is useful in understanding GFCI protection. A ground fault occurs if electricity accidentally finds its way out of its circuit and takes a path to ground through a person or a piece of metal (or often both). The GFCI will sense this and shut off current in a fraction of a second. In most cases, a GFCI can halt the current flow before injury or damage occurs.

GFCI outlets are typically installed in bathrooms, kitchens and garages because the presence of water in these areas increases the possibility for electric shock.

The way to find out if your GFCI is wired correctly is to press the TEST button on its face. This should shut off power to the GFCI outlet and to those outlets connected to it.

If pressing the test button does not cause the GFCI to shut off (but those that follow it do shut off), then it may be wired incorrectly. The wiring may have been installed on the wrong side of the receptacle. The black (hot) wire from the service panel should enter the GFCI on the LINE side. The black wire that feeds power to outlets that follow on the same circuit should exit the GFCI on the LOAD side.

If pressing the TEST button shuts off power to the GFCI and the outlets that follow it on the circuit, then it has been wired correctly, in which case you must press the RESET button to close the circuit again and restore power to the outlet and those outlets on the same circuit.

The drawing shows two GFCI outlets. The outlet on the left is the correct installation. The outlet on the right shows the LINE and LOAD wiring mixed up. To correct the problem, you must cut the power to the circuit at the service panel and switch the wires, or contact a licensed electrician.

ELECTRICAL SAFETY

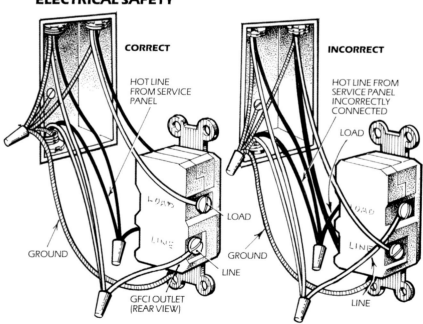

CORRECT

HOT LINE FROM SERVICE PANEL

LOAD

LINE

GROUND

LOAD

LINE

GROUND

GFCI OUTLET (REAR VIEW)

INCORRECT

HOT LINE FROM SERVICE PANEL INCORRECTLY CONNECTED

LOAD

LOAD

LINE

GROUND

LINE

GFCIs Trip Regularly

We recently purchased a newly constructed home. I have continually had a problem with the GFCIs [ground fault circuit interrupters] in the kitchen. The two 20-amp GFCIs closest to the dishwasher trip regularly. It only happens when the dishwasher is running, but not every time the dishwasher is running. I have been told that the devices may be extra sensitive from the moisture associated with the dishwasher. If they are sensitive, can I just replace them? And if moisture from the dishwasher is causing the problem, it seems I have a bigger problem than just tripped GFCIs. Where do I start?

I checked with two dishwasher manufacturers and both agreed that a dishwasher should not be connected to a GFCI. A dishwasher should be installed on a dedicated circuit with a standard 20-amp circuit breaker. In a kitchen, the only outlets that require GFCI protection are those within 6 feet of the sink's outside edge.

It's likely that a tiny leakage current is causing the GFCIs to trip. A GFCI is an electrical safety device designed to cut off current if an imbalance in the current is detected. The current entering the outlet or circuit should be equal to that returning from it. If there is an imbalance in the current flow, it means that current is flowing outside its intended path. A GFCI is designed to detect an imbalance as small as 5 milliamps. If the leakage current varies between 4 and 6 milliamps, it would account for why the GFCIs trip occasionally, but not always, when the dishwasher is running.

It is very unusual for two GFCIs to trip because of a single leakage current. A GFCI, if wired properly, extends protection to outlets that follow it on the same circuit. If a protected outlet also has a GFCI installed (the second device is not needed but could have been inadvertently installed), then a slight leakage current from an appliance plugged into the second GFCI could cause the first device to trip at what appears to be the same time the second device is being tripped. If this is the case in your house, you can remove the second GFCI, but first have an electrician evaluate the circuit.

GFCI, Ground and Neutral

I'm a do-it-yourselfer, and I'd like to know more about GFCI protection and the relationship it has with ground and neutral wires. If I use a small $1/4$-watt electrical circuit tester, I can use the metal junction box or the ground slot as a neutral. However, when doing the same test with a 25-watt incandescent bulb, the GFCI will trip. Can you explain this?

The neutral wire is grounded, but it isn't the grounding wire. The neutral wire is needed to close the circuit. This allows the current in the branch circuit to flow from the electrical service panel through the appliance and back to the grounding bar in the panel.

The grounding wire is a safety feature. It connects to ground components that don't normally carry a current, such as a metal appliance case. If an electrical short develops and the case becomes electrified, the grounding wire directs the current safely to ground rather than shocking the person touching the metal.

The GFCI (ground fault circuit interrupter) is a safety refinement. Under normal conditions, the current is equal in the hot and neutral lines of the branch circuit. The GFCI breaker or GFCI outlet receptacle has a sensing element that monitors the current in both of these lines. If the GFCI senses a current difference between the hot and neutral line as small as 5 milliamps ($5/1000$ of an amp), it automatically trips. The circuit is now opened, and current will stop flowing within a fraction of a second.

Testing the circuit with the hot leg and the grounded junction box or ground slot in the outlet alters the current flow between the hot and neutral legs of the circuit. The GFCI doesn't trip with the $1/4$-watt tester because it draws only 2.1 ma, which is less than the GFCIs 5-ma threshold. The GFCI is tripped when using the 25-watt bulb because that test draws 208 ma. Based on the test you conducted, the GFCI is working properly.

GFCI versus Circuit Breaker

Our house is conventionally wired with 3-wire Romex, a 200-amp service panel and circuit breakers. Over the last 28 years, there have been some electrical modifications to the house. There is a duplex outlet serving a washing machine (a possible lethal shock hazard) that apparently is not protected by a circuit breaker—at least with every breaker switched OFF this outlet is still energized. The wiring of the duplex outlet itself is correct: a black wire, white and ground are present, but somehow it is not connected to a circuit breaker.

If I replace the present outlet with a ground fault circuit interrupter (GFCI), will this give the outlet the same or better protection as a circuit breaker?

A GFCI outlet and a circuit breaker serve different functions and are mutually exclusive. The only thing these devices have in common is that when they trip, they open the circuit. The GFCI's function is to prevent a fatal electric shock. It will trip (open the circuit) in a fraction of a second when there is an imbalance in the current between the "hot" and "neutral" lines that is as small as 5 milliamps.

The function of a circuit breaker is to provide overcurrent protection and thereby prevent a fire. The typi-

cal appliance circuit will trip when the current exceeds 20 amps. However, if there is no imbalance in the circuit, an excessive current will not trip the GFCI circuit.

A little knowledge can be a dangerous thing, as demonstrated by the person who modified your electrical system. All electrical appliances must have overcurrent protection, and the lack of it is a definite fire hazard.

Installing a GFCI outlet for your washing machine for shock protection is worthwhile. However, some GFCIs might trip every time the motor is activated. To prevent this type of nuisance tripping, you should get the hospital grade GFCI outlet rather than the regular commercial grade. It's available at electrical supply stores.

GFCI Questions

My TV and VCR are plugged into an outlet that my kids can reach. Although it has a childproof cover, would I gain additional protection from a ground fault circuit interrupter outlet? Are there disadvantages, other than cost, of having a GFCI breaker in the panel box versus one in an outlet?

A GFCI outlet receptacle certainly provides additional protection against a shock hazard. To do this, the circuit in a GFCI monitors the current in the "hot" and "neutral" lines. Under normal conditions, these two currents are always equal. If the circuit detects a difference between them as little as 5 milliamps, it interrupts the power in as little as one-fortieth of a second. However, childproof covers on an outlet are effective, and it shouldn't be necessary to install a GFCI outlet.

A GFCI receptacle has one advantage over a GFCI installed in a circuit breaker. The GFCI circuit breaker monitors the branch circuit. With it, there is a greater chance of nuisance tripping caused by a buildup of leakage currents due to deteriorated or damaged sections of insulation, multiple splices and moisture accumulation. When a GFCI breaker trips, the entire branch circuit goes out, whereas when a GFCI receptacle trips, it de-energizes just itself or the rest of the branch circuit that follows it, depending on how the electrician has it installed.

ALUMINUM WIRING

Aluminum Wiring Hazard

We bought our home in 1985 and plan to add several new electrical circuits and to install new light fixtures and ceiling fans into already existing circuits. The problem is that all circuits, switches, light fixtures, receptacles and appliance connections were wired with aluminum wiring components.

We know we must replace failed switches and receptacles with ones that are approved for use with aluminum wiring, but we don't know what to do about installing new copper-wired fixtures in existing circuits. A friend warned us that connecting copper to aluminum wiring could create a fire hazard. Is this true?

Aluminum wiring is a potential fire hazard. Between 1965 and 1973 about 1.5 million homes were wired with aluminum, which at the time was approved by the National Electrical Code. Later, it was found that dangerous overheating in 15- and 20-amp branch circuits, at some connections between aluminum wires and outlets, switches, fixtures and appliances, resulted in fires.

Correcting the problem does not require rewiring the house. You can replace switches and outlet receptacles that are unmarked, or marked AL/CU, with devices that are marked CO/ALR. Or, you can use existing switches and outlets provided you attach short copper pigtails to the ends of the aluminum wires and to the devices (aluminum-to-aluminum splices are an exception). It's also necessary to connect light fixtures and appliances with copper pigtails.

Because of the potential for a fire, it is important that copper wire pigtails be attached to aluminum wires with specially designed connectors (Wire-Nuts can be used, however, at copper-to-copper connections, such as at light fixtures). The Consumer Product Safety Commission recommends using a compression-type crimp connector with heat-shrink insulation. These are called Copalum Compression Connectors. They, and the crimping tool used to install them, are made by AMP, Inc. The connections are best left to a

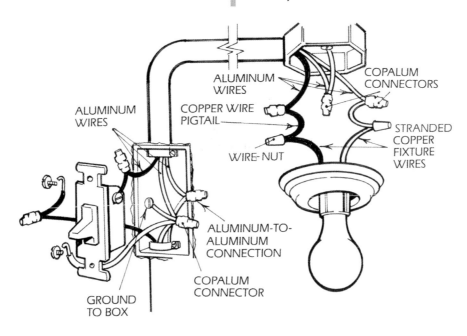

ALUMINUM WIRES

COPPER WIRE PIGTAIL

WIRE-NUT

ALUMINUM WIRES

COPALUM CONNECTORS

STRANDED COPPER FIXTURE WIRES

ALUMINUM-TO-ALUMINUM CONNECTION

COPALUM CONNECTOR

GROUND TO BOX

licensed electrician, who is equipped to work with them and has been trained by AMP, Inc.

Is Aluminum Wiring Safe?

We bought a 13-year-old house that has aluminum wiring throughout. Can you tell me if this wiring is safe?

That depends on whether corrective measures were taken by the previous owner. The U.S. Consumer Product Safety Commission says that houses with aluminum wiring are a potential fire hazard. During the period between 1965 and 1973, because of the shortage and high cost of copper, aluminum was used to wire about 1.5 million homes. Although aluminum wiring was approved by the National Electrical Code, it was later found that there was dangerous overheating in 15- and 20-amp branch circuits at some of the connections between the aluminum wires and outlet receptacles, switches, fixtures and appliances. Anyone who has aluminum wiring should be alert for the following trouble signs:

- Cover plates on outlets or switches that are warm to the touch.
- Sparks, arching or smoke at outlets or switches.
- Strange odors, especially the smell of burning plastic, around outlets and switches.
- Outlets, lights, or entire circuits that don't work.

Even though you may not find any of these trouble signs, the potential for a fire may exist. It can even occur at an outlet that has nothing plugged into it.

Correcting the problem does not require rewiring the entire house. The method recommended by the Consumer Product Safety Commission requires connecting a short piece of copper wire to the end of the aluminum wire by using a special compression-type crimp connector. The copper wire is then used for con-

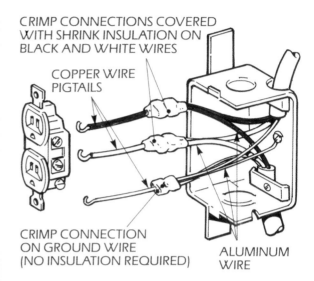

CRIMP CONNECTIONS COVERED WITH SHRINK INSULATION ON BLACK AND WHITE WIRES

COPPER WIRE PIGTAILS

CRIMP CONNECTION ON GROUND WIRE (NO INSULATION REQUIRED)

ALUMINUM WIRE

nection to the outlets, switches, fixtures and appliances. The special connectors, called Copalum Compression Connectors, and the crimping tool are manufactured by AMP Special Industries, Valley Forge, PA 19482.

The corrections described should be made by a licensed electrician who has this equipment. Connecting the copper pigtail to the aluminum wire using a wire nut is not an approved method. You can check your outlets and switches to see if the wire terminations were pigtailed prior to your purchasing the house, but take precautions against electrical shock. If the aluminum wiring in your home has not been corrected, you should take care of this problem as soon as possible.

Aluminum Wiring

I have a home with aluminum wiring. Am I endangering myself and my family by hooking up copper outlets to these wires? If so, how can I remedy the situation and safely connect to the aluminum wires?

Between 1965 and 1973, aluminum wiring was used to install electrical branch circuits in about 1.5 million homes. Subsequent fires in some of those homes were found to be caused by a faulty aluminum wire connection at an outlet. As a result, the U.S. Consumer Product Safety Commission (CPSC) had research conducted that showed that homes using aluminum wires manufactured before 1972 are 55 times more likely to have one or more electrical connections reach "fire hazard conditions" than homes wired with copper.

The solution to this problem is to "pigtail" a length of copper wire to the end of the aluminum wire, and then make the connection with the outlet, switch or other device using the copper wire. Initially, electricians made these connections using a twist-on pressure device called a wire connector. But wire connectors are no longer recommended because a substantial number of them overheated in laboratory tests.

The method for repairing the aluminum wire fire hazard, as recommended by the CPSC and Underwriters Laboratories, is to make the pigtail connection using the Copalum crimp method. This repair uses a crimped connector, which is then covered with heat-shrink insulation.

LIGHT FIXTURES, SWITCHES

Faulty Light Fixture

We recently moved into an apartment building and I think our medicine cabinet in the bathroom is wired wrong. If you turn on the switch, both fluorescent tubes on the sides of the medicine cabinets will light up. When the switch is turned off, both lights glow in the dark. I suspect the switch is wired to the neutral and not the hot wire. What do you think?

It is possible that the switch is wired to the neutral wire, but that alone would not cause the problem. Somehow there is a slight leakage of current to ground that is closing a path for the electrical circuit. This could be caused by deteriorated insulation around the conductor in the junction box or fixture, or even a faulty switch.

Of course, the switch should always be on the hot

wire side of the circuit. If the switch is not faulty, it will open the circuit so that no current flows. When the switch is on the circuit's neutral line, even when it opens the circuit, the fixture will be "hot." If there are stray leakage currents to ground, it would cause the lights to glow.

Time-delay Switch

Please advise me as to where I can purchase a time-delay wall switch. I once had one controlling my carport light, and it became inoperative. I haven't been able to find a replacement anywhere. The switch had a 45 to 60-second delay. It allowed me to turn off the switch, go outside and lock the door, get in my car, and start to drive off before the light went out.

You can buy the switch at electrical supply stores, home centers and hardware stores. There are three buttons on the switch: one to turn the light on, another to turn it off and a third for it to delay turning off. When the delayed-off switch is pressed, the light will remain on for five minutes and then shut off automatically. The switch is for incandescent lights only.

Buzzing Ballast

I have an 8-foot fluorescent light in my laundry room. How can I stop its ballast from buzzing?

The buzzing could be mechanically or electrically generated. Mechanical noise results when the ballast is not securely mounted. To solve this, tighten loose mounting screws and be sure all ballast mounting holes are used. In most cases, the noise is electrical and is the result of vibrations caused by the magnetic flux surrounding the ballast's core.

All ballasts produce some noise, and some ballasts are noisier than others. Although there are no national standards for ballast sound ratings, most companies rate their ballasts.

For example, the ratings of one manufacturer are A, B, C and D. The company's A ballast is for libraries and quiet locations where the desired surrounding noise level is 20 to 24 decibels. The B ballast is for residential locations where the surrounding noise is 25 to 30 decibels, and a C ballast is for a noisier commercial environment, such as a large office with a sound level of about 31 to 36 decibels. The D ballast is for manufacturing facilities with a sound level of 37 to 42 decibels.

The noise from a ballast becomes noticeable when it exceeds the surrounding noise level. That's the problem you have. The ballast in an 8-foot fluorescent light strip has a C rating because an 8-foot strip is normally not used in a residence, except perhaps in a basement work shop. If you find the noise bothersome, have an electrician remotely locate the ballast or buy a fluorescent light intended for residential use.

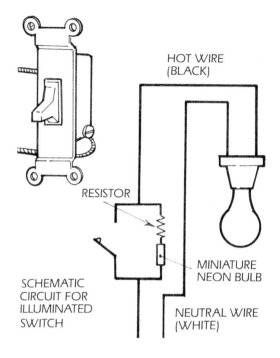

SCHEMATIC CIRCUIT FOR ILLUMINATED SWITCH

HOT WIRE (BLACK)

RESISTOR

MINIATURE NEON BULB

NEUTRAL WIRE (WHITE)

Illuminated Light Switch

I recently bought an illuminated light switch to control the overhead light in my hallway. The illuminated switch lever is very effective in that I can now find the switch in the dark. However, I don't understand how it works. Could you explain it?

Inside the switch is a parallel circuit (see diagram). One leg of the circuit is the switch that controls the ceiling light. The other leg contains a neon glow bulb in series with a resistor.

When the switch is open, current flows through the bulb path. The resistor in this path is sized so the current is only about 1 to 2 milliamps. This is enough to light the bulb, but not the ceiling light. It is also small enough so that if a short occurs in that leg of the circuit, it won't trip a ground-fault circuit

interrupter. A GFCI trips when there is an imbalance in the circuit of 5 milliamps or more.

When the switch is closed (the ceiling light is turned on), that leg of the parallel circuit has little or no resistance compared to the high resistance of the neon bulb leg. Electricity, seeking the path of least resistance, flows through that leg of the switch and lights the ceiling bulb, while the neon bulb dims considerably.

Light-fixture Problem

I recently discovered that the small contact point in the middle of the socket of one of my ceiling-light fixtures no longer protruded enough to make contact with the light bulb. So, I added a drop of solder to the contact point on the light bulb. Everything works fine now, but I wonder, is this safe?

The usual method of dealing with this problem is to raise the position of the small brass tab that makes the electrical connection between the socket and the bulb. To do this, turn off power to the fixture, remove the bulb and carefully insert a small screwdriver into the bottom of the socket. Gently pry up the brass tab on the socket's bottom. Reinsert a new bulb in the socket, turn on the power to the fixture and turn the fixture on. The bulb should light properly.

However, here's how you can check to see if your unorthodox repair will create a problem. Turn the light on and let it stay on for two hours. This should be enough time for the temperature of the bulb to rise and stabilize. Now remove the bulb from the socket, and look at the drop of solder. If the solder shows signs of melting or has changed shape, or if there are signs of arcing on the socket or the bulb, replace the socket or the entire fixture.

Failed Bulbs

For many years I have noticed that the light bulbs in our home burn out every year when cool weather arises in early fall. What causes this, and what can be done to solve the problem?

It's possible that there is an increase in the voltage supplied to your house when and just before you notice the bulbs are failing. A bulb that operates at just a few volts higher than its rated voltage will have a drastically reduced service life. According to General Electric, you ran determine the projected life of a bulb by using the following formula: Actual life = rated life (rated voltage/actual voltage)13.1.

If, for example, you're using a 120-volt bulb, rated to last 750 hours, and the voltage it's running on is 125 volts, then you can expect the bulb to last about 440 hours. You should call your power utility company and have it check the house's electrical supply voltage over a one-year period. Also, ask your neighbors if they've experienced this. If so, an overvoltage is likely.

Multiple Door Chimes with One Button

Please describe how to install a doorbell system activated by one doorbell button that causes two door chimes to operate at the same time. The chimes are located in separate parts of the house. I understand how to install multiple door buttons for one door chime, but that's not my problem.

If you can install a doorbell system with a single chime operated by a single button, then adding another chime is relatively simple. The chimes are connected in a parallel circuit as shown in the diagram. You can run 18-gauge bell wire from the terminals of one chime to the terminals of the second chime.

Most chimes have three terminals which enable them to be activated by two separate buttons—one at the front door and one at the rear door. The center terminal is usually for the transformer connection, and the right and left terminals are for the front and rear doorbell connections.

When wiring the chimes, make sure that you connect the wires to the same terminals on each set. It's also important that you use a transformer of adequate size. A 10-volt transformer works with a single set of chimes, but for a two-chime system, you'll need a 16-volt transformer to do the job.

Faulty Doorbell Wiring

My front doorbell does not ring the chimes. I removed the pushbutton and crossed the wires, but still no chimes. The rear doorbell works, so I switched the wires on the chime unit, front to rear. When I press the rear bell, it rings the chimes as if it came from the front.

I concluded the chimes are working and the transformer must be okay. Is there something I missed?

Check the continuity of the wire that runs from the doorbell button to the chimes and the continuity of the wire that runs from the button to the transformer.

One of these wires probably developed a crack in the conductor due to a kink or sharp bend in the wire. Consequently, there is no longer a continuous circuit.

Check the integrity of each wire with a continuity tester available at electrical supply and hardware stores.

Disconnect the front doorbell wire from the button. Splice a test wire onto the exposed end and run it back to the chimes. Disconnect the bell button wire from, the chimes and touch the probe of the continuity tester to it while clipping the other end to the test wire. The tester's bulb should light. If not, that wire is bad and needs to be replaced. Repeat the test on the wire that runs through to the transformer. Most likely the problem will be solved after replacing one or both wires.

WIRING

Low-voltage Wiring

My neighbor's house was wired in 1955. The electrician installed low voltage wiring for the light switches. Now some of the lights don't work, and my neighbor can't seem to get them fixed. Why was this system used rather than 110-volt wiring?

Low-voltage light switches were installed so that the lights could be controlled from three or more locations. With this system, it is also possible to control all the lights in the house from a master panel. The circuit for this type of system includes a transformer and an electrically operated switch (relay) that is usually mounted near the fixture. In all probability, the problem with the lights is defective relays.

Have your neighbor take one of the relays to an electrical supply house to see if a replacement can be ordered.

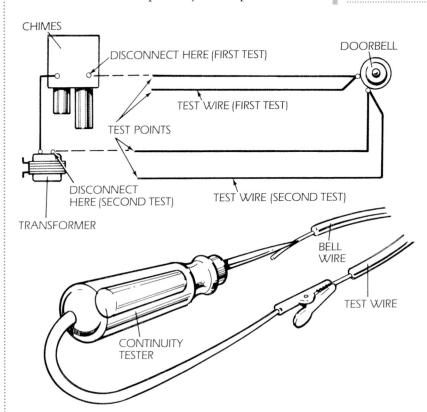

Wire-gauge Question

The general question I have, and I believe most do-it-yourselfers have, is about the mixing of 12-gauge and 14-gauge wires in a home-improvement project. An electrician told me you can go from a 14-gauge to a 12-gauge wire, but you cannot go the other way. I figured that if you connected the 12-gauge wire to one side of an outlet you could safely hook a 14-gauge wire to the other side and extend it to additional outlets. Do you agree?

No, I don't, and it's important to understand why. Wiring the electrical outlet as you suggest is a potential fire hazard. A 12-gauge wire is rated at 20 amps, whereas a 14-gauge wire is rated at 15 amps. All branch circuits lead back to the panel box where the wiring for each branch circuit is protected by either a fuse or a circuit breaker that is rated at the some amp capacity as the branch wire connected to it. When a branch circuit is overloaded, the wiring gets very hot, which causes the fuse to blow or the circuit breaker to trip.

I agree with the electrician who told you that you can go from 14-gauge to 12-gauge wire, but cannot go the other way. If the outlet was wired so that the 14-gauge wire was connected to the panel box, and the 12-gauge wire became overloaded, it would cause the fuse or breaker to open the circuit thus avoiding a fire.

On the other hand, if an outlet was wired as you suggest, with a 12-gauge wire going back to the panel box, and the 14-gauge wire become overloaded, it would heat up but would not cause the circuit breaker or fuse to open the circuit. This could result in a fire.

Generally speaking, an electrical wiring project is not the place to try to save money, especially by using electrical wire that has less current carrying capacity. It's simply easier, and more mechanically sound, to install electrical components that are of high quality and of the same current carrying capacity.

Polarized Plugs

Can I file down the wide tip on a polarized plug without bad effects?

No! Inserting a polarized plug the wrong way around, which becomes physically possible by filing down the wide prong, could cause a shock hazard by making the appliance cabinet electrically live even with the switch OFF. The slots in a polarized receptacle are of different sizes to prevent this. The wide slot is connected to the neutral wire and the narrow slot to the hot wire. The polarized plug ensures that the inlet side of the appliance switch is connected to the hot lead. This keeps components beyond the switch inside the appliance from being electrically hot when the switch is turned OFF.

The two-circuit diagrams show how a polarized circuit can protect you from shock hazard.

If you have nonpolarized outlets and need to plug in a polarized appliance or TV, don't file down the wide prong. Instead, you should replace the nonpolarized outlets with polarized outlets and make sure they have been wired correctly.

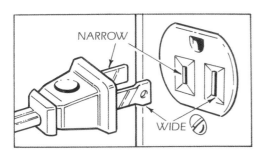

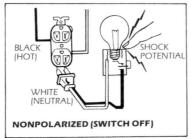

NONPOLARIZED (SWITCH OFF)

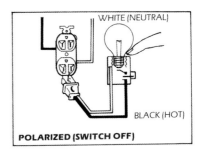

POLARIZED (SWITCH OFF)

Electrical Interference

I have a one-year-old house with a 200-amp service panel grounded with two grounds, one at the water pipe and another to a rod in the backyard. The problem is that there is severe interference on the TVs in the house when wall switches are used, the door bell is rung, the bathroom fan is switched on, or if an electric shaver is used. This happens even though the branch circuits are unrelated. Any suggestions?

This type of problem is often caused by a loose neutral line, possibly at the pole, the meter pan, or in the main breaker panel. Call the local utility company and have them check the connections at these points. The utility company may complete its inspection, but ask that a licensed electrician check the connection at the breaker panel.

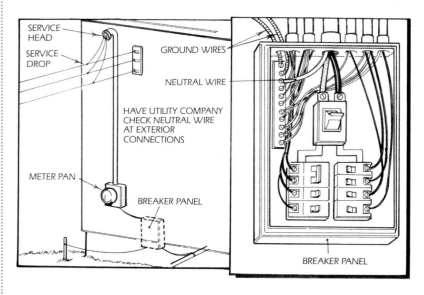

SERVICE HEAD

SERVICE DROP

GROUND WIRES

NEUTRAL WIRE

HAVE UTILITY COMPANY CHECK NEUTRAL WIRE AT EXTERIOR CONNECTIONS

METER PAN

BREAKER PANEL

BREAKER PANEL

12. Plumbing

DRAINAGE/VENTS • SEWAGE-EJECTOR PUMPS
WELLS/PUMPS • MUNICIPAL WATER SUPPLY • PIPES
SEPTIC SYSTEMS

DRAINAGE/VENTS

Fresh-air Vent

There is a large bent, pipe coming out of the ground near my house. Is this pipe connected to my drainage system, and if so, can the roof runoff that flows into the gutters and downspouts be directed to this pipe?

The roof rain runoff can be connected physically, but don't do it. It's usually a violation of the building code and can eventually result in higher taxes. The bent pipe coming out of the ground is the "gooseneck" cover to the house fresh air inlet pipe that is tied into the sanitary drainage system.

In many communities, the house drain line leading to the sewer must have a house trap (see drawing). Its purpose is to prevent noxious gases from circulating back through the plumbing system.

The fresh-air inlet pipe is connected to the house drain on the interior side of the trap. This pipe maintains atmospheric pressure at the trap to ensure complete air movement within the system. The fresh-air inlet pipe can terminate either on the outside surface of the foundation wall where it should be covered with a perforated metal plate, or as a freestanding pipe with a return bend or gooseneck.

By directing roof runoff to the fresh-air vent, you'll be introducing relatively clean water into the sanitary drainage system, and eventually, to the sewage treatment plant.

Each gallon that reaches the treatment plant reduces the plant's capacity to treat sewage. If everyone connected their downspouts to their fresh-air vents, it would result in the need for additional or larger capacity treatment plants, which would in turn, increase taxes.

Backwater Valve

How can I prevent municipal waste-water from backing up into my house's plumbing? On two occasions, it backed up and spilled out of the washing machine's standpipe.

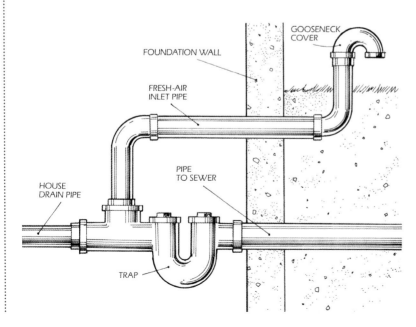

GOOSENECK COVER

FOUNDATION WALL

FRESH-AIR INLET PIPE

PIPE TO SEWER

HOUSE DRAIN PIPE

TRAP

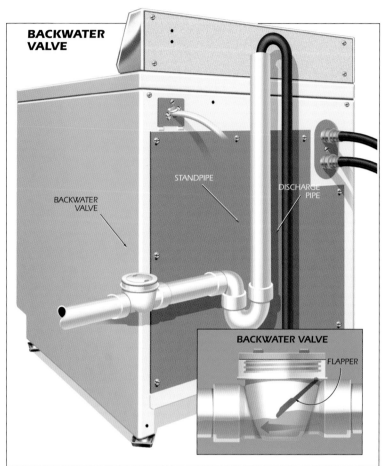

BACKWATER VALVE

BACKWATER VALVE

STANDPIPE

DISCHARGE PIPE

BACKWATER VALVE

FLAPPER

Falling Water Causes Back Pressure

Occasionally the water in my toilet bowl gurgles, and other times it is forced out onto the floor. This doesn't occur when the toilet is flushed but happens at random moments. The toilet works as it should in all other respects. What is the source of the back pressure that is causing this? I live in a condominium apartment on the ground floor of a building several stories high.

A backpressure condition is not uncommon in mid- and high-rise buildings, and it occurs in tall houses that have a bathroom on the top floor. The fixtures in which this occurs are those at the base of a soil stack or where a soil pipe changes its direction abruptly. This is how the back pressure occurs:

A large amount of water, usually from a flushing toilet, will discharge into the soil stack. The water falls through the soil stack (sometimes described as a slug of water), and the air in front of it becomes pressurized because it is unable to slip past it and out the roof vent. Since the sewer pipe slopes, the airspace tapers to a point, which traps the air in the sewer pipe.

The greater the distance that the water falls, the greater the pressure. This is why the problem is found in mid- and high-rise buildings and tall older homes that have had a bathroom added on the top floor or in the attic.

Incidentally, this does not occur when a sink drains into the soil stack. This smaller volume of water tends to cling to the wall of the soil stack and spiral its way down the stack as it falls. It does not pressurize a column of air as does a large volume released from a single flush.

My first thought for correcting the problem is to install a swing-check backwater valve in the main house drain. This would protect the washing machine and any plumbing fixtures in the basement from a reversal of flow caused by an overtaxed public sewer. However, you would need to check with your local building inspector about the plumbing regulations that apply where you live. Some municipalities do not allow backwater valves on the main house drain.

My second thought, and my recommendation, is to install the valve on the branch of the house drain that serves the washing machine, rather than on the main house drain. Most municipalities will allow this installation.

Generally speaking, the products that are available at your local supermarket for unclogging drains will not damage plastic PVC pipes when used according to the directions on the container. If the product can be used on PVC pipes, it will usually say so on the label.

The very nature of household drain-clearing chemicals requires that they be used cautiously. Used incorrectly, they can burn your eyes or skin. None of these products should be used in conjunction with other drain cleaners or general household cleaners containing ammonia. This can result in poisonous gases being produced or a violent eruption of chemicals and wastewater from the drain.

If you find that the drain is still clogged after applying the drain cleaner, I'd recommend that you call a plumber who will use a more powerful chemical drain cleaner or a mechanical snake to remove the obstruction. Look in the Yellow Pages under "Drain and Sewer Cleaning."

Frozen Sewer Line

Our sewer line, from the house to the 1250-gallon septic tank, froze. We had the tank pumped, and the sewer line opened using steam. But it closed up again. To prevent freezing in the future, can the pipe be dug up and wrapped with heating coil and in some way be insulated?

Wrapping the sewer pipe with heat tape and then covering it with a jacketed insulation is an effective way to prevent an ice buildup that clogs the interior of the sewer pipe. There are different types of heat tapes. There are tapes for wet and dry conditions and for metal and plastic pipe. In order to determine the number of coils per foot of pipe, and the length of heat tape needed, you need to know the pipe length, its diameter, and the approximate temperature to which the pipe will be subjected.

Your toilet is probably close to the base of the soil stack, and if the pipe serves a number of bathrooms on the floors above, the pressure developed can cause the air to bubble up through the bowl water causing the gurgling sound. If the pressure is great enough, it can blow the water out of the bowl.

Unclogging PVC Drains

Could you recommend a good household product that I could use to unclog the PVC drain pipes in my house? I've heard so many conflicting claims for these products that choosing one is very confusing. I'm also concerned that these products could damage PVC pipe.

Banging Drainpipes

I have drain lines made of ABS plastic in my home, and the lines are run inside wood frame/drywall partitions. Whenever I run hot water through the line, it makes noise, like someone banging on the walls with a hammer. Can you tell me how to correct it?

Since the noise only occurs when hot water is flowing through the drainpipe, it is probably caused by the expansion of the pipe. Apparently, the pipe is constrained by an inadequately sized opening in the top or sole plate in the wood stud wall. As the pipe expands, it rubs on the sides of the opening, creating the noise. If this is the case, open the section of wall by the pipe, and wedge a piece of polyethylene sheet or plastic milk jug in the opening. This will reduce the friction generated by the rubbing action.

Blocked Vent Stack

Can you explain the workings and purpose of a vent pipe? Some months ago, during a heavy rain, water came pouring into a third-floor room. Our roofer came and said it was not a roof defect, but there might be a break or blockage inside the vent pipe. He filled a plastic beverage bottle with water and inserted it in the top of the vent. We have had no flooding since. However, at times I sense an unfamiliar odor throughout the house. Can you help with this problem?

The odor that you smell is probably from sewer gas, which is the result of the vent stack being blocked. The vent must never be obstructed. Its purpose is to vent sewer gas into the atmosphere and to maintain atmospheric pressure within the drainage system.

Each plumbing fixture in the house, such as a sink, has a bend known as a P-trap as part of its drainage piping. A little water is left in the trap every time water drains from the fixture. The trapped water forms a seal to block sewer gas (which is normally present in the drain line) from entering the house.

When it rains, water enters the opening at the top of the vent stack. Normally this is not a problem. However, if the vent pipe is offset in the attic, and if this offset section is cracked or has a hole, water will drip onto the floor and wet the ceiling below.

If the vent stack is blocked, the sewer gas will rise to the point of the obstruction and then back out to a point where it can escape, such as a crack or hole. It can build up enough pressure to force its way past the water in the trap. In your case, the sewer gas is probably seeping out a crack in the vent stack, and from there the gas makes its way into the attic and seeps into the rest of the house.

I suggest that you look in the attic for the source of the leak. Don't forget to remove the plastic beverage bottle from the vent stack.

Loop Vent

Recently, I had a plumber add a shower stall. The shower is located between the kitchen sink and the bathroom sink. The plumber put a few extra pieces of drainpipe around the shower stall, and told me it was a loop vent and that it's legal. Is it?

Before discussing a loop vent, let's define the purpose of a vent system as part of the drainage/waste system in a house. A vent system consists of a pipe or a series of pipes that are tied into the drainage system. It provides a flow of air within the system, and this equalizes pressure in the drainage system to prevent water from being sucked out of traps (which would let sewer gas into the house).

LOOP VENT

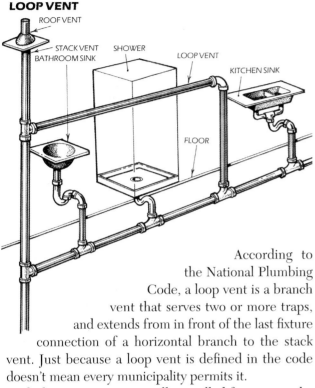

ROOF VENT
STACK VENT
BATHROOM SINK
SHOWER
LOOP VENT
KITCHEN SINK
FLOOR

According to the National Plumbing Code, a loop vent is a branch vent that serves two or more traps, and extends from in front of the last fixture connection of a horizontal branch to the stack vent. Just because a loop vent is defined in the code doesn't mean every municipality permits it.

If a loop vent was originally installed for venting the kitchen and bathroom sinks, all the plumber had to do was tie the shower drain into the branch wasteline. Unless the plumber installed the plumbing originally, or all the pipes are exposed, he would not know the configuration of the venting system. In this case, the shower may not be properly vented.

You should check with your local Building Department. Most municipalities require permits for any plumbing installation. Final approval is granted if the installation complies with the local codes.

Sewage-ejector Pumps

We recently had a lavatory installed in our basement. Because the sewer pipe is overhead, we had to install a sewage ejector pump. We now have a problem whenever the pump goes off. There is a rather loud bang at the end of the ejection process. We have had two plumbers look into the problem, and neither one of them seems to know the cause. Also, we had the floor of the garage dug up and the sewer pipe straightened because we believed that a crooked pipe was the cause. We still hear the banging noise. Any advice?

This problem is fairly common with sewage ejector pumps. The noise is caused by the closing of the check valve. A check valve allows flow in only one direction and is used with ejector systems to prevent sewage from flowing back into the basin after each pumping cycle. This extends the pump's life by preventing it from cycling too frequently. You will not be able to eliminate the noise completely, but you can reduce it.

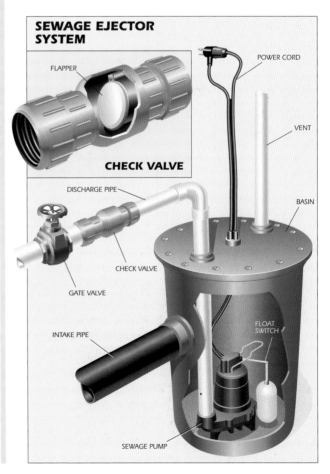

SEWAGE EJECTOR SYSTEM

FLAPPER
POWER CORD
CHECK VALVE
VENT
DISCHARGE PIPE
BASIN
CHECK VALVE
GATE VALVE
INTAKE PIPE
FLOAT SWITCH
SEWAGE PUMP

Where local plumbing codes permit, you can install a section of flexible hose in the piping to decrease vibration and sound. To further reduce the noise transmission, try wrapping a rubberized material around the section of the pipe under the hold-down clamps.

Finally, the valve's installation affects the amount of noise. Some pump companies recommend that the valve be installed in a horizontal position with the hinge facing up. This will prevent solids from lodging on top of the flapper, preventing it from opening again on startup.

Water Hammer in Sewer Ejector

We have a sewage ejector pump for our basement bathroom. The system works well, but makes a loud thud when the pump shuts off after each discharge cycle. There is a 2-inch diameter discharge pipe, and the water column above the check valve is about 8 feet high. Would an air chamber installed in the line just above the check valve help prevent this sound?

Installing an air chamber would help. However, after a while the air would dissolve into the water, and the chamber would become waterlogged. A membrane type of water-hammer arrester is more durable.

I discussed the problem with a technical support engineer at a pump company, who told me that several things can cause the noise.

Quite often the noise is amplified by vibrating pipes. If the horizontal portion of the discharge pipe is loose, secure it to the surrounding lumber with screws and pipe straps. Sewage ejection systems use a check valve to prevent ejected wastewater from flowing back into the system. Some of these valves connect to the surrounding pipe with rubber boots that contribute to vibration noises. Remove the rubber boots, and glue the valve to the pipe.

The engineer said that boring a $3/16$-inch-diameter weep hole in the discharge pipe (about 3 in. above the

water level that activates the pump) will prevent air lock and thereby reduce water hammer. Of course, the weep hole must be located inside the ejector pit.

You also should check that the pump runs in cycles of at least 10 seconds and not for a shorter period—a condition known as short cycling. If the pump is short cycling, call a plumber to evaluate what is wrong. Finally, sewer ejector noises can sometimes be eliminated by installing a second check valve in the vertical discharge pipe about a foot below the horizontal run.

Sewage-ejector Pump

I had a sewage ejector pump installed in my basement when I had a bathroom put in. Now I notice that the pump is running in the middle of the night for no reason. What can be causing this?

A pump manufacturer that I spoke with suspects that a leaky check valve or a hung-up float switch is causing the problem. If debris in the tank causes the float switch to hang up, the pump will continue to run. Also, a leaky check valve can cause liquid to drain back into the tank, which in turn causes the float switch to rise and activate the pump.

The next time you hear the pump running, go down to the pump and listen for water flowing through the discharge pipe. If you hear water flowing, you have a leaky check valve or possibly a leak from the bathroom toilet.

If you do not hear water flowing, then the pump float switch is probably hung up. Open the tank cover to free up the float.

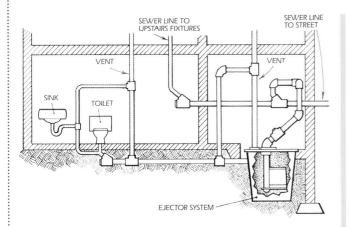

SEWER LINE TO UPSTAIRS FIXTURES
SEWER LINE TO STREET
VENT
VENT
SINK
TOILET
EJECTOR SYSTEM

Up-flush Toilet

I'm interested in installing a half toilet in my basement. However, I don't want to dig up the basement floor for tanks or piping and the sewer line is 5 feet above the floor. An up-flush toilet would be the solution, except that it's not approved by New Jersey. I have been advised that an ejector system is approved for this application, but have found no one who is familiar with such a system. What is an ejector system?

I checked with the New Jersey Code Enforcement Bureau to find out why up-flush toilets are not approved. They said that these units fall into the category of alternative products, which must be tested and approved by a nationally recognized testing agency. Since the up-flush toilets do not carry the seal of approval of an accepted testing agency, New Jersey will not approve of their use.

The only solution to your problem is an ejector system. Unfortunately, this will involve digging up a section of your basement floor.

An ejector system is basically a submersible pump mounted in a tank, which is located below the lowest fixture. The pump lifts the waste to the level of the municipal sewer line. Although the tank is usually below the basement floor slab, it could be buried in a

side yard below the level of the basement floor. This, however, would require a hole through the foundation wall so the tank could be connected to the waste line.

If you're doing the job yourself, don't forget to get a permit from the municipal building department and an inspection and approval by the municipal building and plumbing inspector.

WELLS/PUMPS

Low Well Yield

We bought some land and had a well drilled on it. Good water was found at 130 feet deep, but the flow rate from the 4-inch-diameter well is only 1½ gallons per minute. Are there methods that could be used that would allow us to use this well if we build a home on this property?

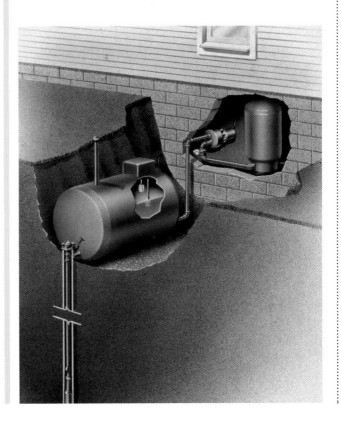

It is feasible to use a well with a yield of 1½ gallons per minute to supply the water requirements for a house. A flow rate of 1½ gallons per minute (gpm) produces more than 2000 gallons of water per day. This quantity of water is more than adequate for most residences.

In order to supply a sufficient quantity of water to satisfy most simultaneous demands, you will need a flow rate of at least 8 gpm. However, a higher rate may be needed depending on the number of fixtures in the house.

A high flow rate (at least 8 gpm) can be accomplished by pumping the well water into a storage tank that is open to the atmosphere and then pumping water from the tank into the house at a greater flow rate. This type of installation is not uncommon and is quite effective. A typical installation consists of a 500-gallon storage tank that is equipped with two sensors—one for high water, which shuts the well pumps and one for low water, which starts the well pump. The high-water sensor can be set at 450 gallons, and the low-water sensor set at 300 gallons. A larger tank may be needed if you have a large family. In order to get the water from the storage tank into the house, you will need a booster pump and a bladder-type pressure tank. The cost for this type of installation depends on the location. Most well-pump contractors and many plumbing contractors can install this type of system.

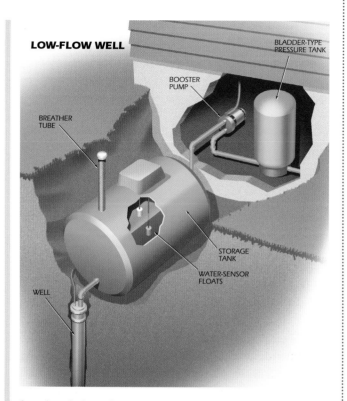

LOW-FLOW WELL

BLADDER-TYPE PRESSURE TANK

BOOSTER PUMP

BREATHER TUBE

STORAGE TANK

WATER-SENSOR FLOATS

WELL

Low-yielding Well

Even before the big drought last year I have had problems with too little water in our well. I can't use our water softener because it runs the well dry at night during the backwash. I would like to install a 300-gallon holding tank in the cellar but am not sure if I should tackle the job myself.

With a low-yielding well you probably also have problems whenever there is a simultaneous draw for water

for the dishwasher and the shower. To resolve your problem you will need more than a pump with a plain holding tank. You will need a storage tank that always maintains an adequate supply of water. This can be done with two sensors—one for high water, which shuts off the well pump, and one for low water, which starts the well pump. This way you will always have an adequate water supply because the storage tank can be filled during those periods of inactivity, such as at night when everyone is sleeping or even when you are drawing water.

In a typical installation of a 500-gallon storage tank, the high-water sensor can be set at 450 gallons, and the low water sensor set at 300 gallons. Depending on your specific water-usage requirements you may want to set the low-water sensor higher.

You will also need a booster pump and a bladder-type pressure tank to get the water from the storage tank into the house. In order to be effective the storage tank must be open to atmosphere. With this type of setup you can achieve a higher sustained flow rate than is possible from a well with a low yield.

I do not consider this to be a typical do-it-yourself project. You may also need a building permit from your municipal building department. I would recommend that you have the local well-pump contractor install the tank, pump, and controls.

Well-pump Cycling

I would like to know if it's normal for a well pump to go on about every 20 to 30 minutes. We bought our house one year ago. It was built in 1962, and it's been a good house otherwise.

Your pump should not be cycling that frequently. There are several causes of this, depending on the type of water system you have. The first thing to check is whether a slow leak somewhere in the plumbing system is causing the well pump to come on. Several likely sources for this are leaky faucets, shower heads or outside spigots. Leaky pipes may also be to blame. While these would generally be easy to spot because of the water damage they would cause, this is not the case if the pipes leak into a crawlspace. In this instance, the increased crawlspace moisture could also cause wood to rot and other problems, such as forming an inviting atmosphere for insects.

Another cause may be a leak in a toilet flapper or ball valve. This can be easily checked by inserting some food coloring or dye in the toilet tank. If the dye shows up in the bowl, you've found the cause, or at least a contributor to it.

If you have a submersible well pump, then another possible cause for the cycling condition is a leak inside the well casing. There may be a fracture in the plastic pipe or possibly a leak in the pump fitting. A leak within the well casing is an expensive problem to correct. Another consideration is whether the storage tank has become waterlogged (no air is left at the top of the tank). If it is, the pump will cycle whenever a small volume of water is drawn or leaks out.

Finally, with a very slow leak, an improperly set or malfunctioning tank-pressure switch will cause the pump to cycle more frequently than it would otherwise.

Well-water Odor

I have a problem with my well water. It smells like rotten hard-boiled eggs. The water is soft and I have no rust problem. I like the water except for its smell. What causes this smell and what can I do to eliminate it?

The rotten-egg odor is caused by hydrogen sulfide gas in the water. Hydrogen sulfate is commonly found in well water and is changed into hydrogen sulfide gas by the activity of non-harmful water-borne bacteria. You can stop the biochemical reaction that forms the gas by killing the bacteria with chlorine. Check with your local health department to determine how much chlorine to add to your well.

Hydrogen sulfate also may be turned into sulfide gas in the water heater (you won't smell the rotten-egg odor at your cold-water tap). In this case, the bacteria react with a magnesium anode in the water heater, and this forms sulfide gas. The anodes are called sacrificial. They corrode instead of the metal heater tank, thus preventing the tank from rusting. You can stop the formation of hydrogen sulfide gas by removing the anode, though this causes the tank to rust out more quickly than it would otherwise. Or, again, you can chlorinate the well.

Oily Well Water

We have a second home in the northern Catskill Mountains. Our well is about 185 feet deep with a submersible pump. Two years ago, the pump broke and had to be rebuilt. Since the installation of the rebuilt pump, our well water has had an oily smell and leaves a film in the toilet. We expected this to disappear with time, but it hasn't.

The person who did the repair claims the oil is from when the pump broke. Can you tell me how to remove this oil from the water?

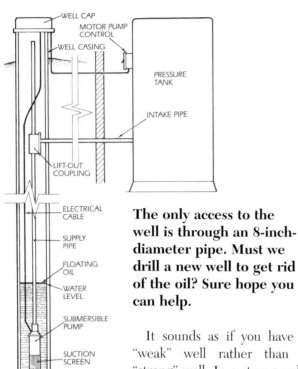

The only access to the well is through an 8-inch-diameter pipe. Must we drill a new well to get rid of the oil? Sure hope you can help.

It sounds as if you have a "weak" well rather than a "strong" well. In a strong well, the water level doesn't rise and fall and, consequently, if oil gets into the well, it floats on top of the water and doesn't get down to the pump. In a weak well, the water level rises and falls frequently, and so the well components become coated with oil.

Since the oil condition has lasted for two years (an excessively long time, when you consider that there is only about half a cup of oil in the pump—under normal use, it should take only a few weeks to flush out the oil) it's possible that the oil is from a leak in a buried fuel oil tank. I suggest you have a water sample tested to determine if it's fuel oil or lubricating oil. If it's fuel oil, you'll have to find and repair the leaking tank.

If its pump oil, you should check further to see if the water is contaminated with PCBs, a known carcinogen. Apparently, some well pumps made prior to the mid-1970s had starting capacitors that were immersed in an oil that contained PCBs. Possibly your old pump was one of these.

Removing oil from the well is difficult and not a do-it-yourself job. A professional pump installer or well driller should do it.

Sweating Water Tank

We have well water. Is there something I can put around the water storage tank to keep it from sweating?

The cold water storage tank is sweating because its surface temperature is below the dew point of the air in that room. This causes the moisture in the air to condense on the tank. You can cover the tank with blanket insulation and then cover the insulation with a vapor barrier of plastic sheeting. Cover all joints in the vapor barrier with plastic tape. An alternative is to buy a water heater blanket, available at home centers, and cut it to fit your tank.

Adding a Tank

I installed a new pump and captive tank in my well-water supply system. My old glass-lined tank is still in good condition. Can I connect the old tank to the new system?

Yes, you can. However, a few words of explanation are necessary. A tank performs two functions. First, it acts as a reservoir from which water can be drawn without having the pump run continuously. Second, it maintains an air cushion that is compressed as the tank is filled. The compressed air is what pushes the water out of the faucet.

Your new captive tank has a fixed membrane that separates the air from the water and prevents the air from dissolving into the water. Your old tank lacks this membrane and works by maintaining a pocket of air over the water. Because the air is in contact with the water, it will, over time, dissolve into the water and the tank will need to be recharged. This may be necessary every 8 to 12 months, depending on water usage.

By connecting the old tank and new system, the amount being held for drawing from the tap will be increased and your pump will start less frequently.

Disinfecting a Well

In a recent column you talked about chlorinating a well to kill the bacteria that cause hydrogen sulfide gas. Could you elaborate on this?

In the area where I live, the following procedure is used to disinfect a well. Mix 1 gallon of liquid chlorine bleach with 4 gallons of clean water. Remove the well cap and pour the solution into the well. Replace the cap.

Open the cold and hot water faucets at the kitchen sink and let the water run until it produces a strong odor of chlorine. Close the faucets and repeat this procedure at every faucet. This allows the chlorine bleach into the faucet body and into the plumbing at each sink. Allow the well and plumbing to stand idle for 24 hours. Then open all the faucets and allow the water to run until the odor of chlorine disappears. Have the water tested to be sure that the chlorine and other contaminants have been removed from the system.

Spurting Well Water

I recently replaced the old galvanized storage tank from my well pumping system with one that has an air bag to separate the water from the air. Since then, whenever I turn on the water faucet, air spurts out of the spout along with the water. Do you know what's wrong?

In all probability, you did not remove the snifter and the drain and Y fitting from your well system when you switched to the new tank.

These valves were installed on well systems that used submersible pumps and a water storage tank with no membrane separating the water and air in the tank. These valves introduce air into the tank to replenish the air lost through turbulence and absorption.

An air cushion at the top of the tank acts like a spring, and as water is pumped into the tank, the air is compressed. The compressed air forces the water from the tank to the spout. Without the air cushion, the tank is waterlogged and the pump performs as if no tank were used, wearing it out prematurely.

In a properly operating system, the pump is stopped by a pressure switch, and the snifter on the check valve opens, allowing air into the pipe. The water slug between the check valve and the drain and Y valve drains into the well. When the pump is activated, the air slug is forced into the tank.

When a storage tank with an air bag or diaphragm is used, the water and air are permanently separated, so it's not necessary to introduce air into the tank. Since you did not remove the snifter, every time the pump is activated a slug of air is sent into the tank. The excess air escapes every time water is drawn.

Remove the snifter and see if that solves the problem. Back out the snifter with a wrench and replace it with a pipe plug. If this doesn't work, remove the drain and Y fitting. However, because the drain and Y fitting are 7 to 20 feet below the well's top, you may need to hire a professional well installer to remove the fitting.

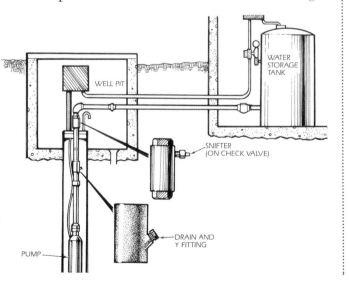

MUNICIPAL WATER SUPPLY

Low Municipal Water Pressure

Our home is higher than the houses in the surrounding area. The municipal water system we're connected to doesn't seem to deliver adequate pressure to match our elevation. Is there a device that can boost our water pressure?

You need three things: a booster pump, a pressure tank, and a pressure switch. There are various-size pumps, but since water pressure for residences is usually between 30 and 50 psi, I recommend purchasing a pump that boosts the pressure 25 psi above street pressure.

The pump should be connected to the inlet water pipe and should discharge into the pressure tank. The tank's outlet should be connected to the house-water distribution pipe. The pressure tank can be quite small and will function only to maintain the pressure. The pump is activated whenever water is drawn.

Depending on your requirements, you could get a larger tank with more storage capacity. The water is stored under pressure until it's needed. As water is drawn from the system, the pressure in the tank decreases. Upon reaching a preset minimum pressure, the switch automatically activates the pump. Also, the pump itself should be mounted on brackets so it doesn't transmit vibrations to the house's plumbing.

Keep in mind that you may need plumbing and electrical permits and inspections to do this job. Contact your local building department before beginning work.

Low Water Pressure

I have a house that is 60 years old. It has ½-inch galvanized iron pipes. The water pressure at the kitchen sink and upstairs bathroom has dropped to about half of what it should be. I had the incoming water checked by the city water department and the pressure was okay. Do I have to replace all the pipes or is there a simpler, less expensive solution?

Although low water flow could be the result of low inlet water pressure, your water department has ruled this out as a possibility in your case. The first thing to check for would be partially clogged faucet aerators at the bathroom and kitchen sink. Unscrew the aerators from the faucets, inspect them, and clean if necessary. If this doesn't help, then the low water flow is probably the result of a constriction in the distribution piping.

Over the years, mineral and corrosive deposits form on the inside of iron pipes. This reduces the effective opening, which, in turn, reduces the flow at the faucets. If the rate of flow is unacceptable, the only solution is to replace the distribution piping and, in many cases, the inlet pipe.

Whistling Line

Our water line has developed a loud whistle whenever a faucet is turned on. This occurs with both hot and cold faucets, indoors or out. Can you help?

A whistling sound in the plumbing system is often caused by water flowing at high velocity past a constriction. Since you hear a whistle when any tap is opened, the problem probably lies in the section of pipe common to all taps—the inlet water service pipe, which contains the water meter and, in many communities, a pressure-reducing valve. I suspect the problem is in either of these two components.

To check this out, get permission from the water company to bypass the water meter. The meter and reducing valve usually have union fittings for easy installation and removal.

If the water meter is the culprit, ask the water company to repair or replace it. If the pressure-reducing valve is the problem, check for sediment in the strainer that could be constricting the flow. Some older reducing valves have brass strainers that disintegrate. If a replacement strainer is not available, replace the reducing valve. Newer valves have stainless-steel strainers that can be cleaned.

You can also try momentarily adjusting the pressure reducer to a higher or lower pressure. This might dislodge sediment buildup.

A Small Plumbing Noise

Shortly after moving into my three-year-old home, I noticed a small plumbing noise. Since then, the noise has increased and often occurs after running water through faucets or flushing the toilet. It happens with both hot and cold water, in the kitchen and the bathroom. The sound is like someone tapping the water pipes with a hammer. Help!

Water hammer could be the culprit. This happens when flowing water is shut off. The water slams into the valve or the end of the plumbing run. You can install an air cushion to stop this.

The water meter may also be to blame, since this occurs when the toilet is flushed (usually not a cause of water hammer). Listen to the meter while water is flowing from a faucet to see if it is the noise source. Also, pipes that are inadequately secured to framing may vibrate against the framing, causing the noise.

Copper Plumbing

I moved into my house when it was finished, 50 years ago, and it is typical of the 500 houses in this development. My house has the original galvanized steel pipes, and last year many neighboring homes were repiped because the original plumbing was in bad shape. I inspected some of the pipes in my house and found they were almost completely plugged by deposits. Plumbers in this area recommend replacing the steel pipe with copper and say it will outlast steel pipe, yet they guarantee their work for only five years. What is your position? Will copper pipe outlast galvanized?

It's really not a question of whether copper tubing is better than galvanized pipe. Both will do the job well. It's a matter of economics. Copper is the dominant material for tubing and pipe in the residential plumbing market. Galvanized pipe is rarely used.

According to the Copper Development Association, copper tubing is used in about 90 percent of residential potable-water systems, with the remainder of the market divided between galvanized steel pipe and plastic pipe and tubing.

The labor costs are greater to repipe with galvanized compared to copper. It takes longer to thread steel pipe joints than to solder copper pipe. Also, a plumber can bend copper tubing for a slight change in direction. Galvanized pipe requires installing a series of fittings to change direction.

Furthermore, my experience with galvanized pipes, as a home and building inspector for the past 25 years, is that it will last about 50 years. Toward the end of its service life, however, waterflow to the fixtures is decreased because the pipe is clogged with deposits.

Copper tubing has a history of providing a service life of more than 60 years. Nevertheless, a small percentage of copper pipe and tubing has experienced premature failure with pin-size holes forming in it.

There are a number of opinions on the cause, such as water composition, dirt particles embedded in the copper tubing during its manufacture, the use of insoluble flux in sweating the pipes, and improper grounding of the house's electrical system.

Incidentally, plumbers generally guarantee the quality of their work, but not the quality of the copper tubing. A five-year guarantee on workmanship is quite good.

Lead-based Solder

I have read quite a bit over the past few years about the dangers of lead in drinking water, especially in connection with soldered pipes. Have newer solders or techniques eliminated this problem?

Although lead-free solder is not a source of contamination, faucets containing lead may pose a problem for homeowners.

Older solder contained equal amounts of tin and lead. Today, there are several varieties of lead-free solder in use, and they may be a blend of two or three metals. A new type consists of about 4 percent copper, about .5 percent silver and the remainder tin. This solder melts at 452 to 464°F, while lead-based solder melts in the 360 to 420° F range. Using the new solders is about the same as the older lead–tin solder, though the higher temperature needed to melt them does discolor the joint slightly. So even if the appearance of a joint soldered with lead-free solder may not be as nice as that of one soldered with the lead–tin type, it is as strong and watertight.

Clogged Pipes Cause Low Flow

The first floor of my house is plumbed with copper pipes, but iron pipes serve the second floor. The water flow from the faucets on the second floor is low. I believe it's due to mineral buildup in the pipes. Does the second-floor plumbing need to be replaced, or can the deposits be chemically dissolved?

A mixture of copper and iron pipes in a plumbing system is fairly common in older homes. Originally, those homes were plumbed with iron pipes. Over the years, as sections of pipes deteriorated, they were replaced with copper pipes. The mixing of copper and iron is a potential problem. A galvanic (electrochemical) action occurs at the joint between the dissimilar metals, which accelerates the corrosion of the joint.

The low water flow is probably the result of a decrease in the inside diameter of the iron pipes. However, I would not recommend flushing the pipes with an acid or other chemical to dissolve the deposits. The deposits may be covering tiny holes in the pipe. Once the deposits are removed, the exposed holes would cause water leaks. Your best bet for long term, trouble-free operation is to replace the iron pipes.

Slow-flowing Pipes

The iron water pipes in our older home have calcium buildup in them, restricting the water flow to a trickle. Is there some chemical means to flush out the pipes? The pipes otherwise seem to be in good condition.

Although the pipes look good from the outside, they are probably deteriorating from the inside out. Besides the calcium buildup, there is probably a heavy rust buildup contributing to the restricted flow.

The only thing that will remove the calcium is acid. However, the acid will also attack the pipes. And if it is not totally flushed out, it could attack you. Also, if you remove the calcium buildup you will probably end up with many leaking fittings, since the calcium seals pitted pipes and leaking joints.

You basically have two choices: don't do anything, or repipe the house. If the water flow is indeed a trickle, I would recommend replacing the pipes.

Copper-plumbing Corrosion

For many years I've noticed a white powdery substance like corrosion around valves and some sweat fittings on my copper pipes. Also, the screws that hold the washers in the valves corrode away. I spoke to a person who claimed the corrosion might be caused by the ground wire—running between my main electrical box and the water line coming from the street—setting off some electrical chemical reaction. Is this true? What do you think I can do to stop the corrosion?

The electrical ground connection is not causing the problem. It sounds as if you have slight leaks around the joints of those fittings and valves. Water oozes out of the pinhole openings in those joints and around the valve stems. It then evaporates and leaves behind the mineral deposits you see.

Usually, the deposits self-seal the leak. However, if the deposits get larger you will have to re-sweat the leaky fittings and repack the valves.

Your washer screws are deteriorating because of the chemical makeup of the water. Home treatment of the water to prevent this isn't practical. The screws should be replaced with monel screws, made of corrosion-resistant nickel-copper alloy, and available at plumbing supply stores.

Noisy Water Pipes

The pipes leading to the exterior faucets on my house rattle loudly sometimes when I water the lawn. The pipes rattle loudest when I partially open the faucet for lower water pressure through the garden hose. I replaced the exterior faucet but this didn't help. Now I think the rattling is because I have copper pipes.

The problem is not because you have copper pipes, but because there is a loose washer somewhere in the water line to the hose bib (the exterior faucet). Water under moderate pressure flowing across a loose washer will cause it to rattle, but under high pressure the washer is pinned down so it doesn't rattle.

I suspect that the loose washer is in the shutoff valve on the pipe leading to the hose bib. It's not necessary to replace the valve, just remove the valve stem and tighten the screw holding the washer in place.

Lead Solder Joints

I've used lead solder for years when installing copper pipe. I've heard that the lead can be hazardous. As I'm just about to build a new home, should I be using plastic pipe?

If plumbing joints are sealed with 50/50 tin–lead solder, it's possible that the water, when not flowing, could pick up trace quantities of lead. As a result, solder containing lead has been prohibited from use in potable water systems.

If your local building code allows the use of plastic pipe for hot and cold potable water, then it's perfectly acceptable for your new home.

PIPES

However, you can use copper pipe by using a lead-free replacement solder such as 95/5 tin-antimony or tin–silver.

You can use the same torches, fluxes, and joint preparation that you've been using with traditional solder. However, the replacement solder melts at a slightly higher temperature, and you'll have to hold the torch over the joint for a slightly longer period of time.

Whereas tin–lead solder (50/50) melts in the range of 361 to 421°F, tin–antimony (95/5) melts in the range of 452 to 464°F, and the melting range of tin–silver (95/5) is between 430 and 473°F.

Tight-fitting Pipes

I have a problem with noise in my house's hot water pipes. When I first turn on the hot water in either of the two bathroom tubs or lavatories, I hear a sound like someone is hitting the pipes in the attic with a hammer. There are about 12 to 15 knocks at 1-second intervals. I have vertical standpipes or air chambers in the attic. I'd appreciate any ideas that would help in getting rid of the noise.

Although the noise sounds like water hammer, it's probably caused by an expansion of the domestic hot water pipes. Water hammer occurs when the water is shut off fast, not when it's turned on.

Unless there's a hot-water recirculation line, and there usually isn't one in a private home, the water temperature in the pipe drops when the water is not turned on at a sink or tub. When the water is turned on, the cool water in the pipe flows out the faucet and is displaced by the hot water produced by the water heater or the tankless coil inside the boiler.

The high-temperature water, in turn, causes the hot water pipe to expand slightly. If the pipe passes through a hole or hanger that is a tight fit, or rubs against floor or wall framing members, the expansion will cause a series of creaking noises, which will stop when the pipe stops expanding.

The problem can be eliminated by allowing for the expansion of the hot-water pipes. If a pipe rubs against wood framing, separate the two with foam rubber. If a pipe hanger is too tight, replace it with one that has a larger diameter.

Sometimes, expansion noises are heard after there has been carpentry or renovation work done in an area that has hot-water pipes. The carpenter may not have allowed enough space between framing members and pipes. If this is the case, you'll know where to look first to correct the problem.

Backflow Prevention

The early weeks of fall are a good time to plant trees and shrubs. Many homeowners water a new planting with a garden hose placed at its base. Water trickles out of the hose and soaks the root ball. However, if a condition called back siphonage (also known as backflow) occurs as the plant is being soaked, water can be drawn back through the hose, contaminating the municipal water supply. The contaminants can range from dirt and silt to fertilizer, herbicides and pesticides.

It can happen when a work crew opens a water main to make repairs, or when water is drawn from a fire hydrant to fight a fire. It also can occur when a vehicle knocks over a fire hydrant in an accident. In these situations, a vacuum or partial vacuum is created in a portion of the municipal water system. Water in the system reverses its flow, and it can draw in contaminated water.

As unlikely as this sounds, many documented cases of backflow contamination have occurred, and the situations described above are only a few of the many possible. Yet the problem is easily prevented by installing a vacuum breaker on the end of the house's exterior faucet. A vacuum breaker permits normal water flow, but backflow causes its diaphragm to close and its vent

BACKFLOW PREVENTION

SET SCREW

SLIDING SLEEVE

DIAPHRAGM

VENT HOLES

VACUUM
BREAKER

holes to open. Water is blocked by the closed diaphragm, and the vent holes let in air to break the vacuum. This way, contaminated water cannot enter the larger water supply system—it is stopped at the vacuum breaker. This devices costs about $20 at plumbing-supply houses.

Warm Water at Cold-water Faucet

When I turn on the cold-water faucet in the kitchen, the water comes out very warm, and it takes a while before it becomes cold enough to drink. The faucet is a two-handle type. However, the water at the bathroom sink comes out cold immediately. It is also a two-handle type. Why does this happen, and what can I do to prevent it?

The piping configuration in your home may be such that the pipe from the water inlet service—where the water is cool because it's underground—to the bathroom is considerably shorter than the length of pipe to the kitchen. The water in the cold-water pipe to the kitchen is picking up heat when the faucet is closed and there is no water movement. Because there is more of this warmed water in a larger pipe, it takes longer to purge the kitchen pipe than the shorter bathroom pipe. Sometimes the heat transfer is the result of the hot- and cold-water pipes being too close. The pipes should be at least 4 to 6 inches apart. You can reduce the heat gain by insulating exposed portions of hot- and cold-water pipes to the kitchen sink.

If the pipes are concealed so that they cannot be insulated, then your best bet for a cool drink is to keep a bottle of water in the refrigerator.

Winterized Hose Bib

The water pipes that run to the hose bibs on the outside of my house are inaccessible. Should I replace my hose bibs with those that shut off the water inside the house (freeze-proof hose bibs)?

Replacing the hose bibs with freeze-proof models would be one approach. The other would be to install an interior shut-off valve that is accessible so the hose bib can be drained.

SEPTIC SYSTEMS

Septic-Tank Cleaning

We moved into a newly constructed home two years ago, and we have a well and septic system. How will we know when it's time to have the septic tank pumped?

A conventional septic system contains two major components—a holding tank and a leaching field. The tank allows anaerobic bacteria to act on the waste and break it down into liquid, sludge, and scum. The sludge accumulates on the bottom of the tank. Then the scum floats on top of the liquid and the liquid flows out of the tank to the leaching field. There, it drains through perforated pipes into gravel-filled trenches. The liquid then percolates into the soil.

If the sludge level gets too high, it will be transported by the liquid into the leaching field. The sludge will clog the pipe perforations, the airspaces between the gravel or crushed stone and even the airspaces between the soil particles below. Over time, the clogged leaching field will become a soggy, odorous mess with lush grass growing over it. In severe instances, waste can back into the house and flow out plumbing fixtures on the house's lowest level.

There are several factors that determine the tank-pumping frequency: the tank's size, the number of people living in the house, whether a garbage disposal is used, and how often laundry is done. Many municipalities recommend a tank be pumped every two to three years. Depending on the factors discussed above, it may need to be done more or less frequently.

A safe approach is to have the tank inspected yearly until it is determined that pumping is required. Once the pumping interval is established, you can follow that until there is a change in water-use patterns that would require the tank to be pumped more or less frequently.

SEPTIC TANK CLEANING

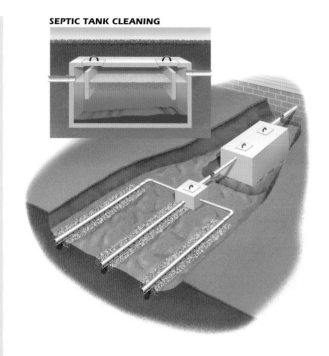

Septic-system Chemicals

Is it really necessary to add yeast or chemicals to a septic tank to maintain balance and trouble-free operation? Thank you in advance for answering this common question.

The most important maintenance item for a septic system is periodic pumping. Most health departments recommend that septic tanks be pumped every two to three years or when the total depth of sludge and scum exceeds one-third of the liquid depth of the tank.

Health officials in the federal government and many state governments have found that the operation of septic tanks is not improved by the addition of chemical compounds, enzymes or septic tank cleaners. In general, the addition of chemicals to septic tanks is not recommended. Some products that claim to "clean" septic tanks contain compounds that may damage the disposal field and actually result in clogging.

The Manual of Septic Tank Practice by the U.S. Department of Health, Education and Welfare, states that although hundreds of these products have been placed on the market (some of which have been marketed using extravagant claims), none have proven to be an advantage in properly controlled tests.

Septic-tank Additives

Our new home has a septic tank and a drain field. Is it a good idea to use a flushable additive to treat the septic system, or is it a waste (no pun intended)?

There are approximately 1200 chemical and biological additives for use in septic systems on the market today. According to the claims in their promotional literature, they are supposed to improve the operation of the septic system and, in some cases, eliminate the need to have the tank pumped. Most of the manufacturers' claims have not been subjected to controlled scientific tests with field data replicated by third-party researchers. In cases where studies have been conducted, there is some debate as to the additives' effectiveness.

Washington State's Department of Health invited all manufacturers to provide evidence from controlled studies that demonstrate positive benefit to septic system performance. None were submitted. Also, according to a document from the North Carolina Cooperative Extension Service, additives do not reduce the need for regular pumping of the septic tank. Furthermore, the document says, some of the additives contain organic chemicals that may damage the drain field or contaminate groundwater and nearby wells.

Generally speaking, state health agencies do not recommend septic system additives. If a system is properly designed, and not abused by dumping garbage or chemicals down toilets, the only maintenance needed is inspection and tank pumping.

Septic Chemicals

I recently purchased a home that has a septic tank and leaching field, and I am looking for information on how to maintain the system. I have also started receiving mailers describing septic system bacteria enhancers that claim, "Put an end to pump outs." Do these products work, or should they be avoided?

First, a good general source of information on septic systems is your state's department of health or extension service.

Because of bacterial action, the wastes that enter the septic tank break down into a liquid and sludge. When the liquid level is high enough, it flows into the field where it leaches into the ground. The sludge settles to the tank's bottom and eventually must be pumped out.

If the system is designed properly and not abused, as from dumping garbage down toilets, the only maintenance needed is periodic inspection and pumping.

Bacterial enhancers are enzymes (catalysts) that speed up the biochemical reaction by which sludge is broken down. Enhancers reduce the rate at which sludge builds up, but don't attack inorganic matter such as sand, grit and ash, which are also present in septic sludge. Consequently, it will always be necessary to pump out the tank. Any company that claims otherwise is obviously using misleading advertising, and its product should not be used.

Whether or not you use enhancers depends on cost effectiveness. If the tank needs to be pumped every four years instead of every two, calculate the weekly or monthly cost of using enhancers during that period and compare it with the cost for pumping. Regardless of what method you use, it's very important that the sludge not build up to where it overflows into the leaching field. The sludge will fill the gaps between the gravel that surround leaching pipes, preventing adequate drainage.

Septic Pumping

We live in a 22-year-old house that has a septic tank that has never been pumped. Assuming that there is excessive sludge in the tank, why hasn't it caused some sort of backup or drainage problem?

The design of a septic system is based on the percolation of the soil in the area of the leaching field and the number of people living in the house (the number of bedrooms in the house is used to estimate this). It is possible that the septic tank was grossly oversized for the number of people living in the house, or perhaps the house was used as a second home and occupied only part of the time. Also, sandy soil in the leaching field is not affected by sludge as quickly as clay-type soil. Many homeowners have never had their septic tank pumped and have not experienced problems. They, and you, have been quite lucky so far. However, these systems could fail at any time.

The most important maintenance procedure for a septic system is a periodic pumping. Most health departments recommend that septic tanks be pumped every two to three years or when the total depth of sludge and scum exceeds one-third the liquid depth of the tank.

The cost for cleaning a septic tank is actually a cheap investment in the life of a house's septic disposal system. If sludge overflows the tank and enters the leaching field, it could cause the system to fail. Digging up and replacing a clogged leaching field costs thousands of dollars.

Septic Seepage Pit

I would appreciate your answer to the proper installation of a seepage pit regarding the crushed stone that is applied around the block liner. How many inches should be applied around the block liner?

For readers who are unfamiliar with seepage pits, a pit is used instead of a leaching field in residential sewage disposal when the lot the house is located on is too steeply sloped to allow building a field. The pit allows effluent to percolate into the ground the way a leaching field does, but it takes up less surface area. Sewage leaving a house settles in a septic tank before it flows into the pit.

The pit's bottom should be filled with 6 to 12 inches of coarse gravel, and the space between the pit liner, and the surrounding soil with 3 to 6 inches of coarse gravel. The specific amount of gravel depends on local codes.

Septic-system Woes

I have a septic system problem. Once a year, I have to dig up my tank cover and clean out the tank. We have a water softener which uses about 20 pounds of salt weekly. We switched soaps, softeners, etc., but nothing helps. Waste foods and grease are kept to a minimum. I would appreciate any suggestions that you could give me on this matter.

Cleaning or pumping out a septic tank once a year really is not that bad. Most health departments recommend that the tank should be cleaned every two to four years, depending on use. Even when a septic system is functioning properly, the tank should be cleaned periodically, or at least inspected for sludge buildup.

Otherwise, the sludge can accumulate to a level where solid wastes will be carried out into the leaching field and eventually clog the voids in the soil and the perforations or open joints in the drain tiles. When this happens the leaching field requires replacement, usually at a cost of at least 10 times the price of a cleaning.

Apparently, the waste water from your water softener regeneration process discharges into your septic system. This is a potential problem, especially if the leaching field is installed in a finely textured clay-type soil. The salt brine in the waste water is not broken down by bacterial action as it passes through the septic tank on its way to the leaching field.

As the salt accumulates in the field it can, and often does, clog the voids in the soil, thereby damaging and shortening the life of the disposal field. The waste water from the water softener should not discharge into your septic system. It should run to a separate dry well, or onto the ground surface away from the leaching field, plants and shrubs. Be sure, too, that this discharge area slopes away from your water well if you have one.

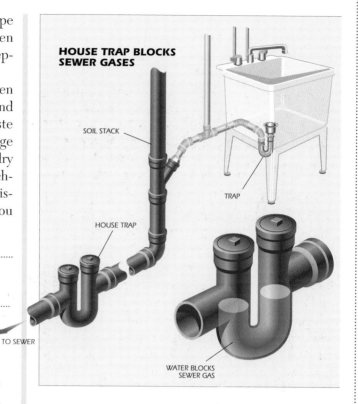

HOUSE TRAP BLOCKS SEWER GASES

SOIL STACK

TRAP

HOUSE TRAP

TO SEWER

WATER BLOCKS SEWER GAS

Septic Odor

About a year ago, we began to get a sewerlike smell in the basement of our house. Our septic tank has never backed up, and it was pumped out about three years ago. I had a septic-tank repairman come out, and he said the system seems to be working right, and he couldn't explain the smell. I checked the roof vents and found no obstructions. Can you suggest a cause for this and a solution?

This problem is common in homes where plumbing fixtures go unused for extended periods of time. More often than not, the fixtures are toilets or utility sinks in the basement, although it can happen with a fixture at any level. Lack of regular use results in the water evaporating from the fixture's trap. The water that's normally present in the trap provides a barrier against sewer gas entering the room.

To correct this, pour some water down the drain to fill the trap. If you do this once a week, you should prevent the problem. In situations where there is a house trap on the main drain/waste pipe leading to the septic or municipal sewer, the condition is greatly reduced. Under normal conditions, there will be water in the house trap.

Septic-tank Location

I have a septic tank but I don't know where it is. It was put in a long time ago and there doesn't seem to be any documentation. How can I find it?

Try checking with the local health department to see if they have a diagram that shows the location of your tank. If they don't have it, then try the following:

Go into the basement and follow the sewer pipe to the foundation wall. This pipe leads to the septic tank which is probably 10 to 30 feet from the wall. Now, go to the area outside where the pipe exits the wall and walk in the general direction of the pipe while striking the ground with a heavy, solid steel, 5-foot crowbar.

SEPTIC SYSTEMS

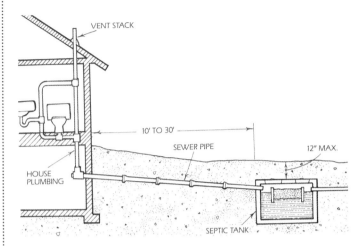

The top of the septic-tank cover (according to proper design criteria) should be 12 inches below grade. Striking the crowbar over solid ground will produce a solid sound. However, when you are over the septic tank, you should hear a hollow sound.

Septic Systems

Do you know of any good books on septic and well-water systems so that a homeowner can be better informed before calling for the repairman?

You can usually get information on septic systems from the department of health or the cooperative extension service in your state. Two books that I refer to regularly are: *Standards for Waste Treatment Works—Individual Household Systems,* published by the New York State Health Department, and *On-Lot Subsurface Sewage Disposal Systems,* published by the Connecticut Cooperative Extension.

The most important maintenance item for a septic system is a periodic pumping. Most health departments recommend that septic tanks be pumped every two to three years or when the total depth of sludge and scum exceeds one-third of the liquid depth of the tank.

If the tank isn't cleaned periodically, the sludge will build up to the point where the solids will be carried into the leaching field and clog the perforations and open joints of the disposal pipes. If this occurs, you'll have to replace the field, which costs thousands of dollars. On the other hand, pumping the tank generally costs hundreds of dollars.

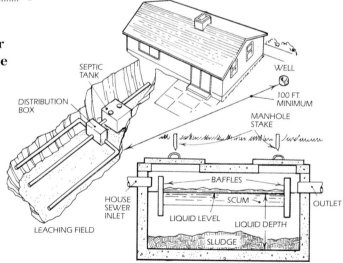

13. Heating System

HOT-WATER SYSTEMS • STEAM SYSTEMS • WARM-AIR SYSTEMS • HEAT PUMP • OIL-FIRED SYSTEMS • GAS-FIRED SYSTEMS • MISCELLANEOUS

HOT-WATER SYSTEMS

Purging Air from Gurgling Baseboards

Lately, we have been hearing a gurgling sound from our hot-water baseboard radiators. Is the noise indicating a dangerous condition?

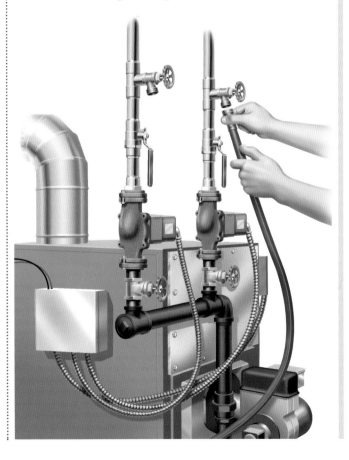

The gurgling sound does not indicate a dangerous condition. It's the result of air that is trapped in the heating system. The air should be purged because it may collect in an area where it will interfere with proper heating.

When a house's radiators are laid out in one loop or a series of loops, there is generally a drain valve and a purge valve located in the return piping of each loop. To purge air from the system, connect a garden hose to the drain valve and run the other end of the hose to a floor drain or utility sink. Next, close the purge valve, and open the manual water-feed valve/pressure-reducing value on the waterline that goes into the boiler. When there is more than one heating loop, purge each separately. The drawing shows two loops that must be purged.

The first step is to open the drain valve and allow water to drain out until all the air has been evacuated. Next, close the drain valve and open the purge valve.

During the purging process, the boiler pressure will drop to below the recommended fill pressure of 12 psi. You must fill the boiler and the entire system to the correct pressure. The problem now is that the fresh water will have dissolved air in it, which will eventually have to be removed. So, repeat the process until nearly all the air is removed.

If there are purge vents on radiators and baseboard convectors, it's also possible to remove air through them. This may be a slower and less convenient way to purge air from the system, especially if there is a large amount of trapped air.

Regardless of which process you use, first fire the boiler for 15 minutes to heat the water to about 180°F. This drives off dissolved gases in the water.

Airbound– Water-logged

Our ranch house has an expanded second floor and is heated with a forced hot-water system. The system is controlled by a single heating zone with four piping circuits. For the most part, it works very efficiently, except for the second floor. I have been advised that to improve the system, I need to periodically close the other three circuits and drain the air-entrained water out of the pipes.

Also, at least twice, the expansion tanks have filled with water and caused the relief valve on the boiler to go off. I have drained the tank, which appears to solve the problem, but why does the expansion tank fill up? Any suggestions will be appreciated.

I assume that your expansion tank is the conventional air-cushion type and not the diaphragm type. If so the two problems you describe are interrelated and can probably be eliminated by installing an air-separator dip tube. The air you are purging from the second-floor heating pipe should be directed into the expansion tank to prevent waterlog.

When boiler water is heated, dissolved air is driven out and rises to the boiler's top. Some air passes into the connection to the expansion tank. However, most of it makes its way into the distribution supply piping. The air rises to the highest point, in your case the second-floor piping, eventually making it airbound.

The boiler and distribution piping in a forced hot-water system are completely filled with water. The boiler water expands when heated, and the increased volume of water flows into the expansion tank, where it dissolves a small amount of air. The on-off cycling of the boiler can eventually deplete the air in the expansion tank causing it to become waterlogged. This will occur if not enough air is directed to the expansion tank from the boiler. A waterlogged expansion tank causes high pressure in the boiler. When water pressure climbs over 30 psi, its relief valve discharges.

The dip tube is a piping arrangement connected to the boiler end of the distribution supply piping. It extends into the boiler so that it is below the air bubbles that accumulate at the boiler's top. Heated water enters the lower end of the dip tube, minimizing air accumulation in the distribution piping. With a dip tube, the air that separates out from the heated water will remain at the boiler's top and also pass into the expansion tank, minimizing the possibility of waterlog. Plumbing and heating supply stores sell dip tubes.

Zero Pressure in Boiler

The pressure gauge on my hot-water boiler does not register over zero pounds of pressure. I am not getting the heat that I used to when the boiler was new three years ago. Do you have any idea what could be causing the problem?

Assuming the pressure gauge is not faulty, a zero-psi reading indicates a compound problem. The boiler has lost its water, and the water make-up valve is defective or improperly set. The valve is set to automatically introduce water when the boiler's pressure drops below 12 psi. Then, if the boiler runs out of water, its pressure will drop to zero or a very low level. This can cause the problem you describe as well as make the heating system noisy. It can also damage the boiler and its circulator.

A boiler can lose its water by any one of several means, some of which are very difficult to detect. Water could leak from a boiler section joint when the boiler is firing. The water will quickly evaporate, and its loss thus goes undetected. If there is a very slight opening in a joint of a heat distribution pipe located in a wall, the water leakage could be slow enough to go unnoticed. However, over an extended period of time,

ZERO-PSI BOILER

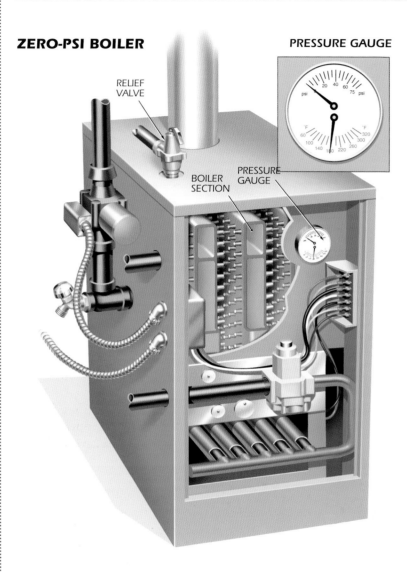

RELIEF VALVE

PRESSURE GAUGE

BOILER SECTION

PRESSURE GAUGE

reduced, the air that is dissolved in the water will tend to surface and rise up to the radiators or baseboard convectors.

You can determine if this is the case by first letting the boiler cool down to ambient temperature and then refilling it to 12 psi. Restart the boiler and periodically check its pressure gauge to see if the system is holding its pressure. If it isn't, then it's time to call a plumber.

Noisy Heating System

I have a baseboard hot-water heating system. A percolating sound comes from the boiler when it fires. A plumbing/heating contractor told me the sound is caused by corrosion inside the boiler. Is this correct, and if so, what can be done to remedy this?

The percolating sound is not caused by a corrosion buildup at the bottom of the boiler, but rather by a mineral-deposit buildup. Here's how it happens: Depending on the hardness of the water, some calcium carbonate and magnesium carbonate (commonly called lime) are in solution in the water when it is initially introduced into the boiler. When the water is heated to more than 140° F, these minerals precipitate out and are deposited at the base of the boiler. The deposits prevent the water in the boiler from coming in contact with the boiler wall, and the boiler wall gets hotter than it would otherwise—that is, without the insulating effect of the deposits. The deposits, in turn, get so hot that they cause a very small amount of water immediately contacting them to flash into steam. Within a fraction of a second, the steam condenses, and this is what causes the percolating sound.

it would reduce the boiler water pressure if the water make-up valve is malfunctioning.

Another possible source for low water pressure is a very slow drip from the boiler's relief valve. Of course, a relief valve that leaks constantly would drop the boiler water pressure, but that would be very noticeable. Check the outlet of the pipe that is connected to the relief valve. It is also possible that the zero pressure is the result of a significant one-time leak that has self-sealed.

The reason you are not getting heat the way you used to is probably that air bubbles are blocking the water flow. When the water pressure in the boiler is

There are a couple of things you can do to try to eliminate the noise. First, try draining the boiler to flush out the deposits. If this is not effective, a plumbing/heating contractor can use chemicals to delime the boiler. However, one boiler manufacturer told me that these deliming compounds are not always effective.

Hydronic Moaning

I have a four-zone, gas-fired, forced-hot-water heating system. The water heater has started making a low moaning noise intermittently, almost a groan. The system is 26 years old and has had minimal maintenance. How can I stop the noise?

It sounds like the bearings are failing on the circulator pump. Over the years, the bearings wear—especially if they have not been adequately lubricated. Worn bearings get hot, overheat, and develop a scraping noise, which is transmitted through the piping and can be heard in some sections of the house. If the sounds are coming from the circulator bearings, replace the circulator.

Circulator pump bearings should be lubricated at least once a year with motor oil, typically sold at hardware stores. Newer circulator pump motors, however, are usually permanently lubricated.

Two-zone Banging

Our raised ranch is heated with hot-water baseboard radiators. We recently had a new gas-fired boiler installed and switched from single- to two-zone heating. With both zone valves working, we get banging sounds, in various places in the house. I ran the system on the downstairs valve alone and then on the upstairs valve alone. In each case, there was no banging in the pipes.

The problem may be caused by a zone valve closing while the circulating pump is still operating. Depending on how your heating system is wired, the thermostat for one zone may be calling for heat when the thermostat in the other zone is satisfied and the valve for that zone has closed. The circulating pump may still be pulling water through the system. When that water, moving at 4 gallons per minute, hits the closed valve, it could make the rattling noise. If the zone valves cannot be repaired, you may have to replace them.

Antifreeze in Pipes

I go to Florida every winter and I am always concerned about the plumbing freezing and bursting in the house I have left unattended. I leave the boiler on and the thermostat set at 55° F. Is it possible to put antifreeze in the heating pipes but leave the boiler operating?

Yes, antifreeze can be used in heating pipes. Also, you should drain the water from the pipes that supply fixtures such as sinks and toilets, and fill sink drain traps and toilet bowls with antifreeze.

However, there are several things that you should do before adding antifreeze to the heating system. First, if the boiler's water feed line doesn't already have a backflow preventer, you should have one of these devices installed. This prevents antifreeze from contaminating the house's water supply and also prevents antifreeze contamination beyond the house. Assuming that your house's water is supplied by a municipal system, a major break in the municipal system can cause enough negative pressure to suck the antifreeze out of a house's plumbing, contaminating the water system beyond the house in the process.

The antifreeze is added to the house's plumbing in proportion to the amount of water in the plumbing. To do this correctly, you have to determine the volume of water in the house's heating pipes.

No-freeze Heating

My home is heated by a hot-water baseboard heating system. In the winter I'm concerned that if the heating system shuts down, the water in the system would freeze. I'd like to know if the system can be winterized.

Yes, you can protect your heating system from freezing. In a hot-water heating system, the furnace, piping, and radiators are completely filled with water all the time. To protect against freezing you should remove some of the water and replace it with ethyl glycol antifreeze, which is available at plumbing-supply stores.

Determine the total volume of water in the heating system. The percentage you remove and replace with antifreeze will depend on the lowest temperature against which you may need protection. For most climates a 50/50 mixture is adequate.

Before introducing the antifreeze, make sure the seals on the circulating pump will not be affected. Check with the manufacturer. Some pumps with seals not meant for antifreeze can be retrofitted with new seals.

Maintain the water level in the system using the manual gate valve rather than the automatic fill valve. If a problem should develop with the fill valve, it could dilute the antifreeze. Also, keeping the gate valve closed will isolate the antifreeze from the potable water supply and prevent the risk of backflow contamination.

Heating-system Replacement

I would like to replace the existing forced-air central heating system in my house using an existing water heater as a source of hot water. Also, what are the disadvantages of central water heating systems?

The main disadvantage of forced-hot-water heating systems applies to homes in areas with freezing temperatures during the winter. In the event of an extended power failure, the distribution piping could burst. Another disadvantage is that the system cannot provide for a central humidification system the way a warm-air heating system can.

The BTU (heat) output of a water heater per hour is considerably less than that needed to adequately heat most homes. This setup is more effective in heating an addition than a whole house. If you use the water heater to supply hot water to baseboard radiators, you will need to keep heating water separate from the domestic (potable) hot water system. That is, you will have to install a second water heater to provide domestic hot water.

Keep in mind that for domestic hot water, the temperature range is from 120° to 140°F (it is usually set at 120°F since 140°F is scalding temperature). For a heating system, the normal operating range is between 180° and 200°F. If you operate the water heater at the lower temperature range (120° to 140°F), you will need more radiators or baseboard convectors per room—at a greater initial cost.

Another reason to separate the potable water from the water used for heating is that during the summer there is no water circulation in the heating system piping. Bacteria grow in this warm, stagnant water .If the two systems are not separate, this water contaminates the potable water when the heating system is turned on.

Radiator Leakage

I acquired a hot-water radiator from a building prior to its demolition. After a cold-water pressure test revealed no leaks, we installed the radiator. Shortly thereafter, we discovered that it leaks about a pint of water a day. Does a slow drip from a hot-water radiator pose any threat to the heating system?

It could. It depends on the hardness of the water and how long the leak exists. In a hot water heating system, the radiator, distribution pipes, and boiler are filled with water. In a properly functioning system, correct pressure is maintained by the pressure-reducing valve, which lets in water when some water leaks out.

However, the introduction of this water leads to mineral deposits on the boiler bottom. Here's how it happens: Calcium carbonate and magnesium carbonate (commonly called lime) are in solution to various degrees in most water supplies and are responsible for the condition known as hard water. These minerals precipitate out of the water when it is heated to more than 140° F. Therefore, the harder the water and the longer the leak exists, the more minerals that are introduced into the system and the greater the amount of deposits on the boiler bottom.

The deposits can insulate the boiler water from the boiler wall. When the water cannot cool the wall, the wall can overheat and crack.

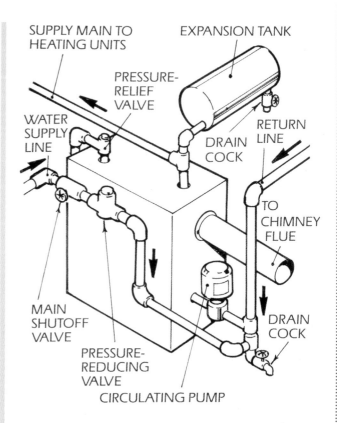

SUPPLY MAIN TO HEATING UNITS

EXPANSION TANK

PRESSURE-RELIEF VALVE

WATER SUPPLY LINE

RETURN LINE

DRAIN COCK

TO CHIMNEY FLUE

MAIN SHUTOFF VALVE

PRESSURE-REDUCING VALVE

CIRCULATING PUMP

DRAIN COCK

Waterlogged Expansion Tank

Our home has an older gas-fired hot-water boiler, with a pressure regulator and a relief valve to maintain the proper water pressure. A 30-gallon cylindrical tank is slung horizontally from the ceiling over the boiler and connected to the main hot-water line.

A small amount of water leaked from the relief valve drain on each heating cycle. The overhead tank was waterlogged, and assuming that it was intended for expansion of the heated water, I pumped air into it. This stopped the leaking from the relief valve but resulted in several airbound radiators. Is this tank an expansion tank or an emergency source of water to the boiler?

The tank is an expansion tank and not an emergency water source. The water-feed line to the boiler, with its pressure reducing valve, will automatically feed water at the required pressure (12–15 psi) whenever makeup water is needed.

In a hot-water heating system the distribution pipes, radiators and boiler are a closed circuit completely filled with water. When the water heats up, it expands.

The additional volume of water flows into the expansion tank where it presses against a cushion of air—or in newer systems, against a diaphragm. Without an expansion tank in the system, expanding water would cause the pressure-relief valve to discharge.

When an expansion tank becomes waterlogged the system acts as if there is no expansion tank, and the relief valve constantly discharges. To correct this, you have to drain the tank. In the process, an air cushion will be reestablished. If you have an older-style tank without a diaphragm, it should be drained completely. A diaphragm-style tank needs to be only about two-thirds drained. If there is a shutoff valve in the line between the tank and boiler, be sure to leave it in a fully open position after draining the tank.

Your airbound radiators are caused by pockets of air trapped in the system. To release this air, you have to bleed the system. All or some of your radiators should have small key-operated valves for this purpose. After the boiler has been operating for several hours, hold a cup under each valve and open it to release the air until you get a steady stream of water.

Waterlogged Tank

The expansion tank on my hot-water-heating-system boiler is barrel shaped and measures about 12 inches long with a 12-inch diameter. The tank has a red plastic insert at its bottom. How can I remove excess water from this tank? I am losing about a gallon a day through the pressure-relief valve. I have replaced the valve and the new one also leaks.

The tank is a modern pressurized diaphragm expansion tank that is used in many hot-water heating systems. The tank's welded steel body encases a rugged, flexible plastic diaphragm that separates the water from the air charge. The tank is pressurized at the factory to 12 psi and the red insert is a cover for the tire valve used to charge the tank. This expansion tank is smaller than conventional non-pressurized tanks because less space is needed for the air.

Unlike conventional tanks, where the air and water are not separated by a diaphragm, this tank cannot be recharged once it becomes waterlogged. It must be replaced.

A discharging pressure-relief valve (assuming the valve isn't faulty) indicates high pressure in the boiler, the cause of which may not be the expansion tank alone. It can be caused by a malfunctioning pressure-reducing valve that steps down incoming water

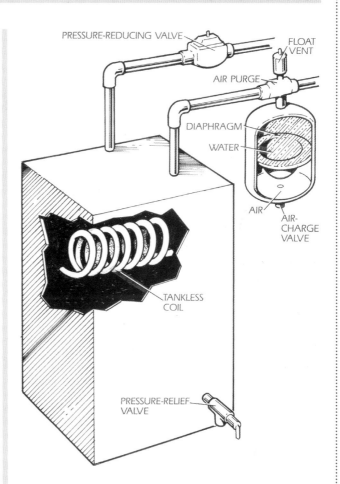

pressure before it enters the boiler.

Also, if the boiler is used to produce domestic hot water, then the cause may be a leak in the tankless coil, which is at a higher pressure than the boiler. Unlike the water in the rest of the boiler, its pressure isn't reduced. A leak in the coil will introduce high-pressure water into the boiler.

A waterlogged expansion tank or malfunctions in either of the other two components can cause a high-pressure buildup in the boiler.

To determine if the tank is waterlogged, remove the red plastic insert and depress the tire valve. If water shoots out, the tank is waterlogged. If air comes out, the tank is not waterlogged; therefore, the problem must be caused by either a faulty pressure-reducing valve or a leak in the tankless coil.

Knocking Baseboards

We have hot-water baseboard heat. When the system is producing heat, there is a constant knocking at both ends of the baseboard units. This knocking also continues while the heat is subsiding. Our house is a split-foyer style with upper and lower levels. It has a two-level heating system. We've consulted several heating contractors, but none can solve the problem.

All piping materials expand and contract with temperature changes. A 50-foot length of copper pipe, any diameter, will expand in length more than $1/2$ inch when the water inside is raised from 70° to 170°F (typical for a baseboard radiator). This expansion can strain joints and cause leaks. It can also make elements bind against radiator covers and jam risers against floor holes, causing noise. Even when provisions are made during installation to absorb this expansion, some noise may still come from the baseboard units.

The noise is probably caused by the heat-distribution pipes or connecting fins rubbing on their support brackets as the pipes expand when the heat is coming up and as they contract when the heat is going down.

This noise can usually be eliminated or reduced by inserting foam-rubber pads between the baseboard support brackets and the connector fins or distribution pipe, whichever is being supported. When inserting the pads, gently lift the heating pipes or fins. If you apply too much pressure, you will strain pipe joints and possibly crack them.

Hot-water Heating

During the next nine months, I will be building a 1850-square-foot home. I like the idea of a hot-water heating system, but living in the Southwest, I find no information available for this. Is there a good reason why this type of heating is not used in my area? I will be cooling the house with a fan, and plan to use lots of insulation.

The only reason that I can think of is one of economics. In many parts of the country, especially in the Sun Belt, central air conditioning is no longer a luxury, it's a necessity. Installing central air conditioning in a home that's heated with hot water is more costly than one heated with warm air where the ducts and fan from the heating system can be used by the air-conditioning system. Also, in many warmer states, heat pumps are used, again because of economics, to supply both heat and air conditioning.

All heating systems have advantages and disadvantages, and you should select the type that best satisfies your needs. If you do not intend to centrally air condition the house, or if installation costs are not a consideration, then by all means, use baseboard hot water to heat the house.

SUPPORT BRACKET

RADIATOR FINS

FOAM RUBBER

STEAM SYSTEMS

Steam-boiler Operation

Could you explain how a steam boiler works? My neighbor has a steam boiler that, for whatever reason, isn't working properly. I'm familiar with hot-water boilers because my house is heated by one. How do these differ from steam boilers?

Steam boilers are quite similar to those used to produce hot water. In some cases, old steam boilers are converted into hot-water boilers. But there are major differences between steam and hot-water heating systems beyond the boiler. In a hot-water system, the boiler and distribution piping are completely filled with water. On the other hand, a steam boiler is only 75 percent full of water. The remaining portion of the boiler, the distribution piping and the radiators are filled with air. They become filled with steam when the system is operating. The two systems use different valves, piping, radiators, safety components and controls.

When a steam boiler is producing insufficient heat, a plumber who is familiar with steam heat needs to give the entire system (the boiler, controls, valves, piping, and radiators) a thorough inspection to determine what is wrong. Sometimes something as simple as a blocked air vent, the removal of pipe insulation, an improperly adjusted control, or even dirt and rust in the water can cause the system to malfunction.

To understand this, it's helpful to know how a steam system operates. Although there are different kinds of steam heat systems, they all heat water and produce steam, as in a kettle. The steam rises from the boiler into the piping and pushes out the air ahead of it as it moves toward the radiators. The air in the pipes and the radiators is pushed out of air valves located on the radiators or on the end of the steam main. When the steam comes in contact with the cool radiator, it condenses and gives up its heat. The water that condensed from the steam flows back through the pipes to the boiler.

Among the problems listed above, a blocked air vent is perhaps the most common. Often the vent is covered with paint. The blocked vent causes the air in the radiator to be compressed by the steam. The compressed air is at a higher pressure than the steam, and this prevents the steam from entering the radiator. The radiator won't heat up, and the system will appear to be not producing steam. This can be corrected by replacing the paint clogged air valve.

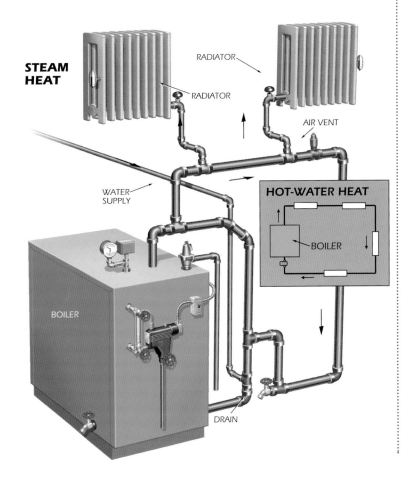

STEAM HEAT

RADIATOR

RADIATOR

AIR VENT

WATER SUPPLY

HOT-WATER HEAT

BOILER

BOILER

DRAIN

Direct Venting

I would like some information on through-the-wall combustion venting for an oil-fired steam boiler. Is this type of vent okay? I would like to vent a new boiler through the wall rather than use the chimney.

Side-wall venting (also known as through-the-wall or direct venting) is becoming quite common with gas-fired boilers and furnaces. When the boiler or furnace is fired by oil, there is a chance that soot could stain the siding adjacent to the vent terminal if the oil burner is not properly tuned or the nozzle is partially clogged by dirt.

Generally speaking, steam boilers are designed as traditional appliances that vent exhaust gases up a chimney. To convert one into a direct-vent appliance, the technician who installs it must fit the boiler with an induced-draft vent kit.

HEATING SYSTEM UPGRADE

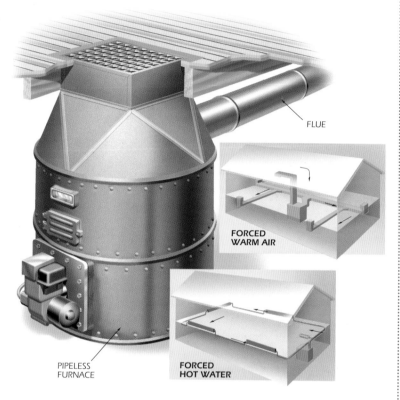

FLUE

FORCED
WARM AIR

PIPELESS
FURNACE

FORCED
HOT WATER

Furnace Upgrade

My husband and I recently purchased a two-bedroom country ranch, built in 1946. The house is in very good condition, but it still has the original monsterlike furnace, which dispenses all its heat through a grille on the main floor. The furnace's thermostat is in the master bedroom, and the door to the room must remain open for the thermostat to sense the house's air temperature. Can you give us any advice about a new heating system?

The heating system in your house is not a central heating system in the modern sense, although I have seen many in operation in vacation homes and cottages. It's called a pipeless furnace, and the warm air it produces simply flows through the large floor register into the room above. It does not use any other ducts to distribute warm air throughout the house, as does a modern forced warm-air system, a simplified view of which is shown in the illustration above.

Since you have a ranch house with a basement, installing a modern central heating system will be relatively easy, especially if the basement is unfinished. The two most common central heating options are a forced warm-air system or a forced hot-water system. The former heats air with a furnace, the latter heats water with a boiler. Both heating systems can use natural gas, propane, or fuel oil. Ask your heating contractor about the advantages and disadvantages of each and which system would be the most economical to install in your area.

Warm-air Heat

I am thinking of buying a house that has warm-air heat. Could you tell me some of the advantages and disadvantages of this type of heating system?

In a warm-air heating system, the air is heated in a furnace that is usually oil or gas fired, although it can be heated by electric-resistance coils. The heated air travels via supply ducts and enters the room through wall or floor registers or ceiling diffusers. Cooler air is displaced by heated air and travels through return ducts back to the furnace.

There are a number of advantages to this system. The air can easily be filtered and humidified or cooled to provide air conditioning.

Also, if there is an extended power failure during the winter, the heating ducts are not vulnerable to freezing and bursting, as are heating pipes.

A disadvantage of a warm-air, gas-fired, or oil-fired system is that, in the rare event of a defective heat exchanger, poisonous carbon monoxide from the exhaust gas can mix with the circulating air and be distributed around the house. This also introduces excessive moisture (from the combustion process), in the form of condensation, into the house.

When the system is off, no heat enters the house, unlike the residual heat that is released from hot-water radiators. Also, a hot-air system has a central air filter that needs to be regularly cleaned or replaced.

Cycling Furnace

The burners on my furnace cycle on and off four to five times before the thermostat shuts them off. In its present condition, the comfort level is excellent with no perceptible overshooting or undershooting of the thermostat setting. I would like to know if this condition is normal.

The condition is not normal, and there are several different malfunctions that can cause it. Normally, the burners fire until the thermostat is satisfied and shuts them off. A malfunction in the furnace's high limit temperature control can cause rapid cycling. The limit switch is a safety control that prevents furnace from overheating. It is activated at a temperature that would not be reached during normal operation. Constant activation of the limit control can damage it and also can cause problems with the furnace's ignition control, relays, and gas valve.

Another possible cause is a low flame or an incorrect low flame signal from the furnace's flame sensor. The sensor turns off the furnace's gas valve if it detects a low flame or if the burner is not lit when it should be.

A malfunctioning flame roll-out switch also can cause rapid cycling. Flame roll-

ROOM PARTITIONS

REGISTER

RETURN AIR TO FURNACE

FILTER

FURNACE

HEAT EXCHANGER

WARM-AIR DUCT

out occurs when the furnace has insufficient combustion air. The flame rolls outside the furnace cabinet seeking combustion air. The switch is designed to turn the gas off if it detects a roll-out condition. Since manufacturers have different designs for their operational and safety controls, you should have the problem corrected by a factory authorized technician.

Furnace Runs in Spurts

We have a relatively new house with a gas furnace. When the furnace comes on, it runs just a minute or so and then shuts off. And, it continues to do this until the house is warm. Could this be a faulty thermostat? Any help would be appreciated.

I doubt that the thermostat is faulty. More likely, the problem lies with either one or both of the following: The fan control is set improperly or the temperature-limit control is set incorrectly. Also, the problem may not be a matter of setting—one or both of the parts may be malfunctioning.

Since you didn't say whether it's the burner or the fan that cycles, you should check the temperature settings of both. The purpose of the fan control is to prevent the fan from circulating cool air. The fan control is a temperature-sensitive switch that turns the blower on and off at preset temperatures. It is independent of the thermostat. When the thermostat calls for heat, the burners should fire. The fan should not operate. If it does, the fan control is either faulty or needs adjustment. After the heat exchanger warms up to a temperature of about 110° to 120°F, then the fan should start to operate. When the temperature setting on the ther-

mostat is reached, it will shut off the burner but not the fan. The fan will continue to operate until the temperature in the heat exchanger drops to about 85°F.

The high-temperature-limit control shuts off the burner when the heat exchanger reaches 175°F (it should not shut off the burners prematurely), and the fan should begin to operate before the burners are shut off by the limit control. If the fan comes on at or near the 175°F shutoff point, air discharging from the registers will be uncomfortably warm. In that case, the fan's temperature control probably needs adjustment or the control is malfunctioning.

Limit and Blower Controls

Can you explain the operation of the limit and blower controls for a warm-air furnace? I have conflicting advice from my heating contractors regarding the controls' settings.

The blower control prevents the fan from circulating cool air around the house. It is a temperature-sensitive switch that turns the blower on and off at preset temperatures. It operates independently of the thermostat. The blower's on/off setting depends on what you find comfortable, although it is often set between 110° and 120°F. When the heat exchanger reaches that temperature, the fan operates. When the temperature setting on the thermostat is reached, the thermostat shuts off the burner but not the fan. The fan operates until the heat exchanger temperature drops to about 85°F.

The purpose of the high-temperature limit control is to prevent damage to the heat exchanger by shutting off the burner. The limit control is usually set at about 175°F.

Slow-heating Furnace

I have a problem with my forced-warm-air furnace. When the furnace ignites, it stays on for about two minutes and then goes off before the blower comes on. The blower stays on for two or three minutes and then goes off. The blower and the furnace don't seem to work together. The air from the registers doesn't get very warm, and it takes forever to heat the house. I would appreciate any help you can give me.

There are two basic controls in a forced warm air furnace—fan control and a high temperature limit control. The fan control is a temperature-sensitive switch that turns the blower on and off at preset air temperatures. It usually starts the fan when temperature in the plenum reaches 110° to 120°F, and shuts the fan off at about 85°F.

The high-temperature limit control is also a temperature-sensitive switch that shuts off the burner at a preset temperature, usually about 175°F. For proper operation, the fan should start before the burner shuts off.

Based on your description, it sounds as if one or both of the controllers are either out of adjustment of faulty. Check the temperature settings on the controllers. If they are not set for the proper temperatures, reset them. If they are properly set, they will need to be repaired or replaced.

Whistling Duct

I recently moved into a new condominium consisting of a street-level apartment and an unfinished walk-out basement. The condo is heated by a forced warm-air furnace. After having the basement finished and heated by a second zone connected to the furnace, I became aware of a problem. There was a whistling sound coming from the ductwork. Each heating zone is controlled by a motorized damper. When the heat was turned on for the street level (and off for the finished basement), there was a whistling sound. When the heat for the basement was turned on, the whistling stopped. Do you have any idea what could cause this?

The problem is probably a defective installation of the damper for the basement zone. If the damper is closed, and the heat is activated for the street level, high-velocity air will rush past open joints around the damper's perimeter and cause a whistling sound.

To correct the problem, it will be necessary to cut a hole in the duct, and then seal the joints around the damper with metal-faced tape. The hole in the duct should then be patched with sheetmetal, and the joints sealed with tape.

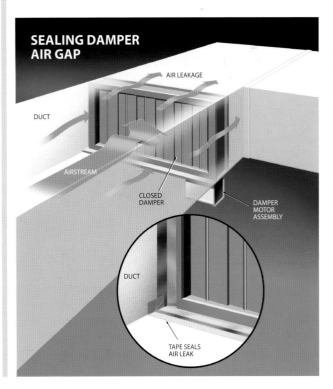

SEALING DAMPER AIR GAP

AIR LEAKAGE

DUCT

AIRSTREAM

CLOSED DAMPER

DAMPER MOTOR ASSEMBLY

DUCT

TAPE SEALS AIR LEAK

Taped Return Register

My husband and I have a running argument about a register in our bedroom. We don't heat our bedroom in the winter, so he covers up the return air register with tape. I say that the tape stops the removal of cold air from our room. He says that if the hot air is shut off, the return air register should also be shut off. Which is the correct answer?

I agree with you, but it depends on what your husband is trying to accomplish. By sealing off the return register, he is isolating your bedroom from the central heating system, especially if the bedroom door is kept closed. Unless you keep a window open in your bedroom for air circulation, sealing the return register will result in the room's air becoming stagnant. This lack of air exchange can cause some discomfort, especially if someone smokes in the room.

By not sealing the return register, there will be air circulation within the room. The return air will create a negative pressure in the bedroom. In turn, this causes the air in the house to migrate into the room through the door opening or under the door if the door is closed.

The extent to which sealing the return register will reduce your fuel consumption is difficult to determine. I would assume that the effect is relatively minor.

Noisy Furnace Fan

I have a problem with my forced-air furnace. The cold-air return chamber is right under the unit in the front. When the heater is on and the fan is running, there is a high fan noise. Is there any way that I can dampen this noise?

Fan noise is usually caused by vibrations of the fan–motor assembly or, if the return is inadequately sized, by air rushing through the grille. Sometimes it's a combination of both.

Since the louvers on a grille will reduce the effective opening, try removing the grille. If there is no change in noise level, you've eliminated the return size as a problem. If the noise level decreases, you will need a larger return or a grille with less air resistance. Next, check the fan and motor assembly. Resecure any loose sections.

You can decrease the noise level from the fan and motor by lining the blower compartment with a matted fiberglass acoustic liner. The liner comes in $1/2$- and 1-inch thickness and can be glued to the sides of the compartment. The acoustic liner is manufactured by insulation companies and should be available through heating/air-conditioning contractors.

You can further reduce the noise (if space permits) by covering the return opening with a lined duct that has an open end several feet away from the fan compartment.

Popping Furnace

We bought a home which has an LP gas furnace. The furnace is about 10 years old and had never been hooked up by a service dealer. The previous owner of our house said he mainly used the fireplace for heat. However, when he used the furnace he hooked up a small tank of LP gas and the furnace worked fine.

When we moved into the house we had the furnace checked out and hooked up by a service company. The unit worked properly when the man left, but in the evening there were loud popping sounds and the gas burned back by the air controls. The serviceman returned and adjusted the unit, but the problem keeps coming back. The serviceman has been here seven times and has changed the furnace control and checked the regulator

valve outside on the tank. We still hear popping sounds by the burner. The company that hooked up the unit is trying to come up with an answer. Do you have any ideas?

The popping sound that you hear is the gas flame being extinguished at the various ports on the burner. I called the manufacturer who said that your LP gas furnace was originally manufactured to burn natural gas and was retrofitted in the field to burn bottled gas.

Two types of conversion kits are available, one for cast-iron burners and one for steel burners. It is possible that the wrong kit was used for your furnace. If it was, the gas ports in the burners would clog easily, causing the flame at the clogged ports to extinguish and pop.

Even if the correct kit was used, it is possible that the burner gas ports are partially blocked and should be cleaned. It is important that the person servicing your furnace be familiar with your equipment. Check with the manufacturer of your furnace for a qualified serviceman. If he can't correct the problem, he can get help from the manufacturer's technical field staff.

Secondary Heat

The high-efficiency oil furnace in my home is vented to a flue in the chimney. When the furnace is operating, the vent pipe gets hot and heat is lost up the flue. I've seen flue fans that are temperature regulated and mounted to a wood-stove chimney exhaust pipe. The fan turns on at a preset temperature, blowing across smaller pipes in the unit and sending a large amount of heat into the room. Would such a unit work on the exhaust flue of my oil furnaces?

A heat-recovery system using a fan to blow air across the flue pipe could work, but it's generally not recommended. Such a system could extract too much heat from the flue gases, which contain water vapor, sulfur dioxide and nitric oxide (and other nitrogen compounds). If the gases get cool enough, the water vapor will condense out of the flue gas and combine with the sulfur and nitrogen compounds. This condensate can cause corrosion in the flue pipe or in the furnace. Also, a low exhaust-gas temperature could affect the draft over the firebox and decrease the efficiency of the oil furnace.

Controlling Airflow

I need to know how to control the airflow in my house's ducts. The third floor receives inadequate airflow for heating and cooling. I've noticed what appears to be a control on the duct near the furnace. How does this affect airflow?

Forced-air systems have (or should have) balancing dampers in the branch ducts that lead to the rooms. The dampers are near the plenum to which the ducts are connected. A handle on the outside of the duct controls the damper. When the handle is parallel to the duct, the damper is open, and air flows through the duct. When the handle is perpendicular to the duct, the damper is closed.

You should probably have a contractor evaluate your house's duct system and its airflow, because even a small adjustment to the dampers can have a major effect on air movement. A 1/4-inch damper movement can increase or decrease airflow in the range of 100 to 200 cubic feet per minute.

One thing that a homeowner can do to improve system performance and comfort is to run the system's fan continuously. Continuous air circulation mixes the air between rooms and between floors. This evens out temperature variations, and prolongs the life of the fan motor because it no longer has to constantly stop and start.

Also, talk to a contractor about the duct configuration in your house. You should consider separate heating and cooling zones for each floor. Separate zones also conserve energy.

Furnace Duct Layout

Our forced-air heating system has a cold-air return in the ceiling about 8 feet from the room's two heat-supply ceiling registers. It seems that the heated air is drawn back through the cold air return. Is this a poor design?

I don't have sufficient information to say whether the heating system's duct arrangement is a poor design. The design may have been a compromise, based on cost, room layout, and the installation of an air-conditioning system. I would say that it is not an optimal or energy-efficient design. It contributes to stratified air layers in the room and short cycling of the heated air, rather than achieving uniform and comfortable air distribution. As you well know, heated air rises, so supply registers should be located in the floor or at the base of the walls. Furthermore, they should be located on exterior walls, preferably below windows. The return grille should be located in the ceiling or high up on the wall opposite the supply registers.

For air conditioning, the supply registers should be located in the ceiling or high up on the wall, since cool air drops. The return should be on or near the floor.

Air-intake Benefit

What are the benefits of having a fresh-air intake on a gas-fired, forced-air furnace? I'm especially interested in the health benefits of this feature. We live in a tight house in a very cold region of the country.

If the house is very tight and there is minimal outside-air infiltration, a fresh-air intake is necessary to ensure that the gas furnace works at peak efficiency. Specifically, a fresh-air intake helps ensure complete combustion of the gas. It also reduces the chance that carbon monoxide from the furnace will reenter the house, a phenomenon in tight houses known as backdrafting.

A fresh-air intake is still a good idea in houses that are not very airtight. In this case, the fresh-air intake conserves energy by ensuring that combustion in the furnace does not rely on the air in the house, which has already been heated.

The cycle works this way: There is generally a negative pressure in a house. When heated air is used in the combustion process, it goes up the chimney as waste gas. This increases the negative pressure since the air in the house is being used to burn the gas. In a house that's not very tight, this causes more cold-air infiltration. This cold air must be heated, which uses up more energy. An air intake uses unheated air to ensure combustion. Also, the negative pressure in the house does not increase because the air inside the house is not used for combustion.

For most homeowners, health problems resulting from indoor air contaminants are minimal. The typical American house built within the past 30 years has an average infiltration rate of 0.4 to 1 air change per hour. The air exchange rate for the living areas of the house should be 0.35 air change per hour.

However, some of the newer, tighter houses have an air exchange rate of 0.1 air change per hour. Without ventilation, this could result in stagnant indoor air, containing more than the usual amount of pollutants and bacteria. To prevent this, you must increase the

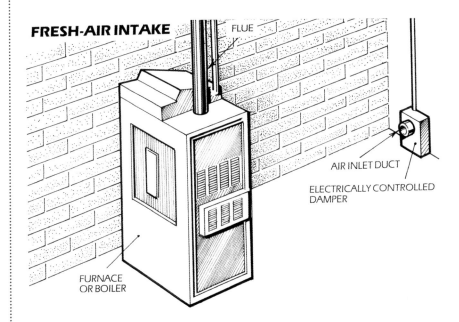

FRESH-AIR INTAKE · FLUE

AIR INLET DUCT

ELECTRICALLY CONTROLLED DAMPER

FURNACE OR BOILER

through the draft control and into the furnace room. This can happen when the house is tightly sealed. Air leaves the house but is not replaced, and this results in lower pressure inside the house than outside. This negative pressure causes backdrafting (the movement of exhaust gases back into the house).

Once the soot is in the furnace room, fine particles of it will be picked up by the air currents in the house. The particles drift into a room and are drawn into a heating-system return register. From there, they find their way to the suction side of the air filter in the furnace room.

amount of ventilation. Consult your local Yellow Pages for heating/cooling and ventilation contractors.

Sooty Air Filter

When I changed the air filters in my father's oil-fired, warm-air furnace, they were coated with black soot on the side from which the air enters the blower. There is soot on the basement floor around the furnace and on the furnace itself. It is not a heavy accumulation, but noticeable. How is the soot getting on the suction side of the filters? I did not see soot around the supply registers in the rooms.

You should have an oil-service technician check your system to determine the cause of the problem. The soot can be caused by a combination of things. For example, there may be inadequate combustion air. This causes inefficient oil burning and results in a soot buildup in the combustion chamber and exhaust stack.

Soot can find its way into living spaces when exhaust gas is pulled back down the soot coated chimney,

Filter Replacement

For the past few years, I've noticed that the filter on my forced-air gas furnace does not need to be cleaned or changed from one season to the next. I seem to get ample heat and cool air through the registers. Prior to this, I changed filters every month or so. Can you tell me what would cause this condition?

Sometimes, a change in living habits can make the air inside a house cleaner. For example, perhaps you stopped using a fireplace that occasionally let in a puff of smoke.

It's also possible that the filter doesn't fit properly and there are openings between the filter and the frame. Consequently, the dusty return air is bypassing the filter. If this is the case, you should check the evaporator coil (you imply in your letter that you have central air conditioning). The coil may be coated with a heavy layer of dust, which will decrease the air-conditioner's efficiency.

Contaminated Indoor Air

I purchased a condo two years ago, and its heating system is horrible. I sneeze and cough most of the time when it comes on. I dust, and just a few minutes later there is this fine dust everywhere. I had the furnace serviced and cleaned professionally. I put wet towels over duct outlets to catch the dust. I use a humidifier, which helps moisten the dry air, and I replace its air filters regularly. Your thoughts on this problem would be appreciated.

I suspect that dirty heating ducts, not the furnace, are the main source of your problem. Warm-air ducts have been shown to be a collecting area for a variety of contaminants, such as mold, fungi, bacteria, and very small particles of dust. It's likely that the ducts have never been cleaned. A thorough cleaning of the ducts is generally recommended prior to moving into a new house. Even when there is a furnace filter, some dust gets through the filter and into the airstream. A portion of it is deposited on the duct wall linings, while the remainder is discharged into the rooms. Over the years, the dust and other contaminants can build up on the duct walls to the point at which problems, like the ones you describe, will occur.

A company that is certified by the National Air Duct Cleaners Association (NADCA) should professionally clean the ducts. Certified NADCA members must pass an examination to demonstrate their knowledge of heating/ventilation/air-conditioning systems and duct cleaning methods. They also must sign a code of ethics.

Another source of airborne irritants is a dirty humidifier. A humidifier should be cleaned before the start of the heating season.

Duct Dust

Several years ago, we had a furnace installed and started getting a great amount of dust every time it was on. We had the same company that installed it check it, but were assured everything was okay. After a couple of years, we asked someone else to check it out. It was discovered that a return-air duct had been laid over the dirt floor in the crawlspace with no cap, and all the dust was being brought into the house. Other than taking the ducts down to remove the dirt, is there anything we can do? It's a year-round problem because the central air-conditioning uses the same ducts.

All warm-air furnace systems must have a filter in the return duct to trap dirt and dust in the air stream. The filter prevents airborne particles from getting into the air supplied to the rooms. There could be several ways that the airstream that enters the room picks up particles. The filter may be dirty, it may not be the correct size or it may not have been installed properly so particles slip past it. Or, you may not even have a filter. If none of the above cause the problem, you might have an electrostatic filter installed in the return duct. These filters are effective in removing dust and dirt from the air.

Although the end cap on the return duct in the crawlspace was finally installed, you should also seal the joints between the duct sections with duct tape. Open joints could be a source of dust entering the airstream. Also, the portion of the return duct in the crawl space should be insulated as an energy-saving measure.

Electronic Air Cleaners

How effective are the electronic air-cleaning systems that can be retrofitted to an existing furnace? Are they easily installed?

There are a number of companies that manufacture electronic air cleaners that are installed in conjunction with a warm air furnace. Honeywell, Trane and White Rodgers, to name just three. These devices can be quite effective at removing pollutants such as dust, pollen, animal hair, lint, smoke, and cooking grease. According to published literature, these filters will remove 90 percent of airborne particles, while a standard 1-inch-thick disposable furnace filter removes less than 10 percent. Electronic air cleaners use an electrical charge to capture particles as they enter the furnace. Return air entering the filter passes through a powerful electrical field where the particles in the airstream receive a positive electrical charge. As the charged particles move through the filter, they are attracted by and hurled against the negatively charged collection plates. They remain there until you clean the plates. Do this by removing them and washing with water.

Installing an electronic furnace filtration system is a do-it-yourself job only if you have experience with sheetmetal fabrication. Otherwise, an experienced heating/air-conditioning contractor should install it. An existing warm-air heating system will need modification to the sheetmetal ducting, and it may need additional electrical wiring, depending on the kind of system you install.

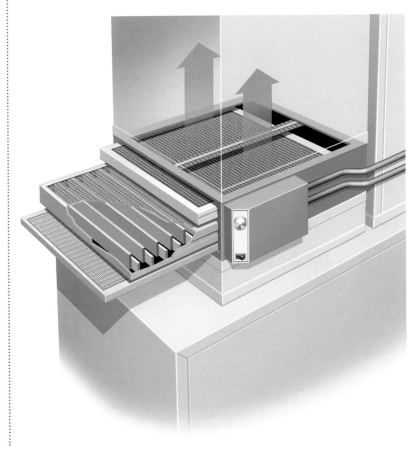

Humidifier Sediment

My hot-air furnace system is equipped with a power humidifier. The water in our area is considered hard and no softener is used in my system. The plastic basin and carousel of the humidifier must be cleaned regularly and an attempt is always made to do so. Despite these efforts, a hard sediment forms on the basin and carousel which is extremely difficult to remove. Do you have any suggestions as to how I can remove the hardened sediment? Also, how can I prevent it from recurring in the future without resorting to adding a complete water-softener setup to my entire water system?

The hardened sediment that forms on the basin, carousel and float assembly in your humidifier can be softened by soaking those items in a 50 percent solution of white vinegar and warm water for about half an hour. The sediment can then be removed and the items cleaned. The components should be rinsed thoroughly before being replaced, otherwise a vinegar odor will be picked up by the circulating air.

There are commercially prepared liquids and/or tablets that can be used to help clean the humidifier and minimize the extent to which the sediment becomes hard. Humidifier tablets are available at plumbing supply stores.

Unless you use distilled water, which is hardly practical, deposits will always build up in your humidifier. Even the use of a water softener will not eliminate this problem. In fact, in many cases it could aggravate it. Quite often the salts in softened water build up in the humidifier faster than the minerals in unsoftened water. Regular cleaning and tune-ups during the heating season is always necessary for effective operation of any humidifier. The frequency of cleaning, however, depends on the mineral content of the water.

Don't forget to clean your humidifier at the end of the heating season before the sediment becomes too hard, rather than at the beginning of' the season, when it's as hard as a rock.

HEAT PUMP

Heat-pump Advice

In 1992 we built a new home and had an electric heat pump installed with an electric furnace backup. A father-and-son heating company installed the heat pump. The father says that when the temperature drops to 20° F or below we should cut off the heat pump and go with the furnace backup, because the heat pump is only providing about 20 percent of the heat. The son says to leave the heat pump alone. Which one is right?

This is not a question of right or wrong. The answer depends on your preference about comfort relative to efficiency.

In those parts of the country where the heating requirements of a house are greater than the cooling requirements, heat pumps are not sized to satisfy the building's design heating load. If they were, they would be oversized for cooling and the cooling performance would be poor. Instead, on a cold winter day the heat pump works together with supplemental heat to provide the required amount of heating for the house. On an overall basis, this is quite efficient because even at low temperatures, the energy needed to operate the heat pump is normally less than the heat energy delivered from the pump.

Although efficient, this system may not be as comfortable as you desire. With an outdoor air temperature of 20°F, the temperature of the warm air discharging from the wall registers will be 80° to 85°F. Although this is enough to heat the room, it is less than body temperature. That is why the air being discharged from the registers will feel very much like a cool breeze. On the other hand, if you switch the thermostat to the emergency heat setting, you will shut down the heat pump and activate the electric furnace backup. This increases the temperature of the air being discharged to about 110° to 120°F

Open-Loop Heat Pumps

I have a well that I do not use. Can I use this well in a ground-source heat-pump system? What are the limitations of such a system?

For readers unfamiliar with them, most ground-source heat pumps use a closed continuous loop of plastic pipe filled with a heat-exchange liquid (usually water and antifreeze solution). The liquid absorbs heat from the earth in the winter and dissipates heat to the earth during the summer. These systems may be configured

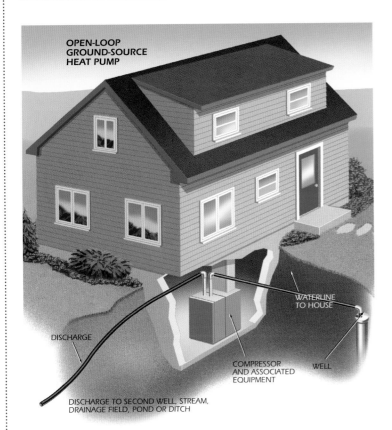

OPEN-LOOP GROUND-SOURCE HEAT PUMP

DISCHARGE

DISCHARGE TO SECOND WELL, STREAM, DRAINAGE FIELD, POND OR DITCH

COMPRESSOR AND ASSOCIATED EQUIPMENT

WATERLINE TO HOUSE

WELL

Heat Pump versus Furnace

My husband and I are getting ready to build our first home (1500 to 1900 square feet with two stories). We are trying to decide on a heating/cooling system. Many people recommend a heat pump because it is energy efficient and pays for itself. Others say a heat pump needs a backup system because it can't heat a home when the outside air falls below 32°F. Still others suggest a gas furnace. What are your thoughts on this confusing subject?

Heat pumps are most effective in the Sun Belt. They lose their economic advantage in colder climates when the outside temperature drops near or below freezing because their auxiliary electric heaters kick in. Considering the Northern location of your home, if the decision is based on economics, I'd go with the heat pump.

in a variety of ways, such as horizontally or vertically, but they are always known as closed-loop types.

The system you are referring to is known as an open loop. This type uses water from a well, lake or stream as the heat-exchange liquid. The water is discharged into a stream, ditch or pond. Unless the water can be recovered, the system is considered wasteful of water. A rule of thumb is that at least 3 gallons of water per minute are needed for each ton of cooling capacity. The open loop should only be used where there is an adequately large supply of water to handle the needs of the heat pump running continuously during extremely hot or cold weather.

Whether or not your well yield is adequate for an open-loop system needs to be determined prior to deciding which system to use. You would also need to check with your local building department to see if there are ordinances that pertain to this type of system.

Cool Heat Pump

I recently moved into a condominium with a heat pump for central heating. When the unit is operating, I feel cool air coming from the register, although the house is warm. Is something wrong with the heat pump?

The heat pump is working correctly. It heats air to about 90°F, which feels cool because it is lower than body temperature, but is warm enough to heat the room. The air discharged from an oil- or gas-fired furnace is about 110° to 130°F. To compensate for the lower temperature, heat pumps require a high air-flow rate, and may create a drafty feeling.

OIL-FIRED SYSTEMS

Old Oil Tanks Are Wallet Busters

My 550-gallon heating tank is 28 years old and is buried in front of my house. I've heard that a leaking tank can result in wallet-busting costs, especially if contaminated soil is the result. What is the typical life span of a residential heating-oil tank? My oil-consumption records contain no evidence of leakage.

What you heard is correct. The cost for excavation, carting and dumping contaminated soil in an EPA approved site can cost thousands of dollars. Depending on how much soil is contaminated, this can cost in the tens of thousands of dollars and could exceed $100,000. The fact that your oil-consumption records do not indicate a leak is misleading. Your tank could have started leaking at the rate of a gallon a year 15 years ago. A leak that small could easily be overlooked. Nevertheless, there now would be 15 gallons of oil contaminating the soil.

A friend of mine who tests buried oil tanks for leaks has found tanks that leak after only five years.Steel tanks usually last 15 to 20 years, although many tanks last longer.Since your tank is 28 years old, you should have it tested for leakage. You might have to make several phone calls to get information on buried oil tanks. First, call your state's environmental agency.

Erratic Oil Burner

We had a new nozzle and filter installed in our old off furnace. After being shut off overnight, it fires for 20 seconds and then stops for 12 seconds. Then it fires like it should until the thermostat causes it to shut off. If the burner is shut down for five or six hours, you have to press the reset button to get it going again. The technician who installed it is stumped. What is the remedy for this?

The problem may be caused by an air leak in the oil-supply line (the suction line). This can cause the formation of an air pocket by the burner nozzle, which could account for a 5- or 6-second delay. After running for a while, another air pocket could cause the safety control to shut down the burner. And when the burner sits idle, the air leak could cause the fuel pump to lose its prime. In this case, when the burner is turned on but doesn't fire within the lockout time on the primary control, it will be shut down by the safety control.

I checked with burner manufacturer R. W. Beckett Corporation. The company says this is often caused by an air leak around the filter or a valve stem, or in a compression fitting, a flare fitting or the nozzle adapter. Another cause is a loose nozzle. The company points out that compression fittings on an oil-supply line are liable to create an air leak. They should be replaced with flare fittings. To find the leak, an oil-heat service technician will have to pressurize the oil-supply line.

Oil Burner Tune-up

I have an oil-fired furnace, and I change the oil filter every other year. However, if an oil-fired furnace ignites properly and gets an adequate amount of fuel oil when it ignites, is there any point in installing a new fuel nozzle as well? Can anything else go wrong with a fuel nozzle in addition to dirt accumulation in its orifice? What is your opinion concerning the need for replacing a fuel nozzle on an oil-fired furnace?

Homeowners often make the mistake of visually inspecting their furnace and concluding that it works properly and efficiently. But even an oil furnace that

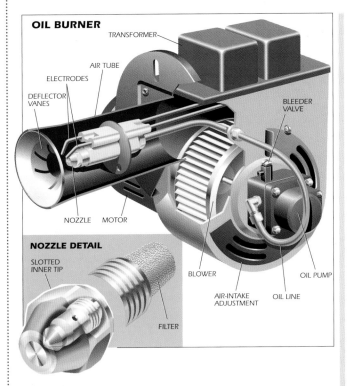

OIL BURNER

TRANSFORMER

AIR TUBE

ELECTRODES

DEFLECTOR VANES

BLEEDER VALVE

NOZZLE MOTOR

BLOWER

OIL PUMP

AIR-INTAKE ADJUSTMENT OIL LINE

FILTER

NOZZLE DETAIL

SLOTTED INNER TIP

oil to swirl as it is forced through the orifice. These slots can become coated with sludge and gum, and in some cases it is practically impossible to see whether the slots are thoroughly clean without using a microscope.

A nozzle orifice is polished to a glasslike finish. If it becomes clogged and you try to clean it with a wire or pin, you can ruin the finish. This will cause streaks in the spray. Considering that a nozzle costs about $10, I would certainly recommend replacing it every time the oil burner is tuned up.

Oil Burner Causes Interference

We have a vexing problem with radio and TV interference that occurs immediately prior to our oil burner firing. Our oil burner technician cannot figure it out. The burner and the thermostats appear to be okay.

The interference is being generated by a problem with the oil burner's ignition system. Oil burners have ignition transformers that step up the house voltage to 10,000 volts. This high voltage is needed to ignite the oil spray.

There are two ignition springs connected to the 10,000-volt transformer output. These springs are joined on the other end to two electrode rods, which cause the arc that ignites the oil. There must be good, clean contact between the rods and the springs. Sometimes one spring will touch the rod, and the other doesn't quite make contact. This causes arcing at that point, instead of at the end of the rod. The arcing can cause radio and television interference.

Your oil-burner technician must make sure that there is no arcing at the springs and that there are no cracked sections in the porcelain insulators around the electrodes.

functions without generating an odor, puff-back or rumbling noises may not be operating efficiently. There are a number of variables that affect an oil burner's efficiency. Some of these variables are draft over the firebox, stack temperature, gas concentration, smokiness of the exhaust, and the proper oil-spray pattern for the shape of the combustion chamber. Because of the complexity in addressing all these things, tuning up an oil burner is generally a job for a professional technician trained in oil-burner repair and maintenance.

The oil-burner nozzle, which directly affects the spray pattern, can have problems aside from dirt clogging its orifice. A nozzle can get overheated as a result of an improper burner-tube setting or back pressure in the combustion chamber, and it can develop gum and sludge formations on the inside and outside. Nozzles have precisely ground slots in the inner tip, which cause

Oil Burner Location

I have steam heat in my home, and when the boiler was converted from coal to oil, the burner was placed at the bottom of the boiler and fired directly into the firebox. However, a service technician later determined that this setup wore out the firebox and said it would be more energy efficient to fill the box with pellets and relocate the burner to the middle of the boiler. The system now produces steam in half the time it took before, but I am concerned about this new arrangement. The burner fires directly on the cast-iron water jacket, as opposed to the firebox. Will this reduce the life of the boiler?

There are two good reasons why the oil burner should not fire on the cast-iron jacket: First, it results in incomplete combustion. To burn completely, oil must be burned so that the tiny particles vaporize. The large amount of water in the jacket absorbs the flame's heat. This reduction in flame temperature results in dirty and inefficient combustion and the formation of soot (unburned carbon). Second, the flame brushing against the cool metal sides of the jacket results in uneven heating of the metal surface. This induces stresses and strains that shorten boiler life.

You should be able to correct the problem by adjusting the tilt of the burner so it fires away from the jacket or by changing the burner nozzle.

Oil Burner's Late Ignition

I have an oil-fired heating system. It runs fine for several cycles, and then its Beckett burner has a late ignition. It then runs fine until the next time. The repair technician has checked the pump pressure and replaced the transformer and protector relay to no avail.

Aside from those you mentioned, several things can cause this problem. There may be air in the oil lines. The air forms bubbles that migrate to the nozzle and cause a delay in ignition. Also, it's possible that the electrodes are burned back slightly or are not gapped correctly. Also, the electrodes could be positioned too high above the nozzle.

A service technician should check these items. If there is still a problem, he or she should call the R. W. Beckett Corporation while still on the job. Beckett will work with the technician over the phone to diagnose and repair the problem.

GAS-FIRED SYSTEMS

Exhaust Stack Damper

My gas-fired furnace has no stack damper on its flue like my old boiler had. My heating service company told me a gas furnace has a damper built inside.I looked at my furnace and didn't see anything that looks like a damper. Could they be wrong?

Boilers with stack dampers are oil fired, not gas fired. With an oil-fired furnace or boiler, it's important for the oil burner to have the proper draft over the firebox. To ensure this, a draft regulator is mounted in the exhaust stack near the boiler or furnace. The regulator is a small swinging damper that opens when the burner is firing.

Gas fired boilers require a small but steady draft supplied by a draft diverter (also known as a draft diverter hood). The diverter is usually located on the flue above the boiler. If the boiler has a pilot light rather than an electronic ignition, the draft diverter prevents a downdraft from blowing out the pilot light flame. It does this by diverting the draft through the open bottom of the hood.

One final point, a gas-fired furnace uses an unobtrusive draft opening in its sheetmetal casing.

Flooded Chimney Flue

I have a new two-story house with a gas hot-water heater and a gas furnace. When the exhaust gases hit the clay chimney flue tiles, they condense. On the average, I collect half a gallon of water a day from the chimney clean-out pit. If I don't siphon the pit, the water seeps into the basement

The furnace has an electronic flue damper. The vent connector is 25 feet long, on a sloped horizontal run from the furnace to the brick chimney. The water heater is connected after the damper to the same vent. The chimney does not have a flue cap, but I get little or no rain in the flue that I can notice.

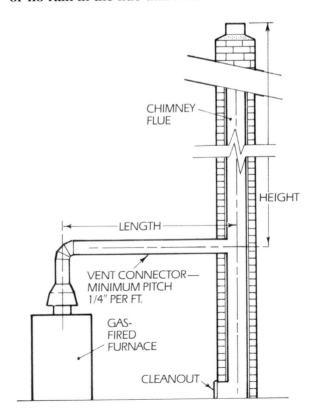

Would capping the flue create a better draft? Would disconnecting the flue damper help? Or running a smaller diameter metal flue within the existing chimney?

The problem is not with the chimney but with the length of the vent connector running to the chimney. The exhaust gases give up an excessive amount of heat as they travel through the long pipe. You might think that's good, because you can heat the basement for nothing, since the exhaust gases are going up the chimney anyway.

However, water vapor is one of the byproducts of combustion in a furnace. By giving up their heat, the exhaust gases are cooled down to a temperature that is at or near their dew point. In other words, the temperature is too low to allow the gases to make it up the chimney without the water vapor condensing.

As a matter of fact, this is a problem often encountered after installing one of the various kinds of heat recovery devices on the vent connector with the hope of saving energy. Too much heat is extracted from the exhaust gases and the products of combustion condense before venting to the atmosphere.

I don't understand why your furnace is 28 feet away from the chimney. You didn't mention the height of your chimney, but it is probably no higher than 25 feet. According to the American Gas Association and the National Fire Protection Association, the horizontal run of a vent connector to a natural draft chimney should not be more than 75 percent of the height of the chimney above the connector. Unless there is a design criteria with your system that you didn't mention, your vent connector is too long for a natural draft, and requires a draft inducing fan.

To minimize the condensation, you might try insulating the vent connector or using insulated pipe. This will reduce the heat loss of the exhaust and keep its temperature above the dew point.

Through-the-Wall Combustion Venting

The exhaust for our furnace goes out the side of the house instead of up a chimney. I'm concerned that this is not a proper installation, and that furnace exhaust will filter back into the house through nearby windows.

In recent years, the venting of exhaust gas through a sidewall has become more common. Most of the heating appliances that use sidewall venting are designed for this. In other cases, heating appliances that were

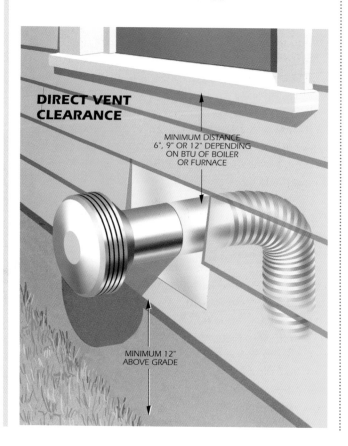

DIRECT VENT CLEARANCE

MINIMUM DISTANCE 6", 9" OR 12" DEPENDING ON BTU OF BOILER OR FURNACE

MINIMUM 12" ABOVE GRADE

not designed for sidewall venting are retrofitted with a fan that permits them to vent through a sidewall. This is known as a retrofit induced-draft installation. This can be tricky, and it's best handled by a contractor experienced with this kind of work. An inexperienced contractor might not match the fan capacity to the exhaust output of the heating appliance. This can cause the appliance to function inefficiently and produce excessive carbon monoxide and, in the case of oil-fired appliances, soot.

Another type of sidewall venting appliance is known as a direct vent. It does not use a fan, and it takes its combustion air from outside the house—an induced draft appliance uses interior air for combustion. A direct-vent appliance may have a vent terminal located relatively close to doors and windows. The vent terminal for an induced draft appliance must be located farther from doors and windows than that of a direct-vent appliance.

There isn't room here to show every configuration of vent terminal. The example shown is a direct-vent terminal for a gas-fired appliance. The drawing is based on specifications in the National Fuel Gas Code.

It's a reasonably involved process to determine whether an appliance complies with the manufacturer's installation guidelines and local or national building codes. If the appliance was installed by a reputable and licensed contractor, a gas utility or a reputable and licensed fuel company, then it's almost certain the installation is okay.

If you wanted to go through the trouble of checking, you would need to look at the appliance's data plate to gather the manufacturer's name and location, the BTU rating of the appliance and its model and serial numbers. You would need to contact the manufacturer to find out if the appliance was designed for sidewall venting or is a retrofit system. The manufacturer might also want to know the location of the vent terminal relative to the ground, doors and windows. As you can see, gathering this information and checking with the manufacturer would be time consuming.

Finally, you can play it safe by installing carbon monoxide detectors in living areas near the vent terminal.

Thermostat's Heat Anticipator

In a previous column, I answered a question concerning a furnace that cycles on and off every other minute. My answer discussed the cause in terms of the furnace's fan and temperature-limit controls. However, I inadvertently left out the thermostat's heat anticipator, which is more likely the cause of the problem.

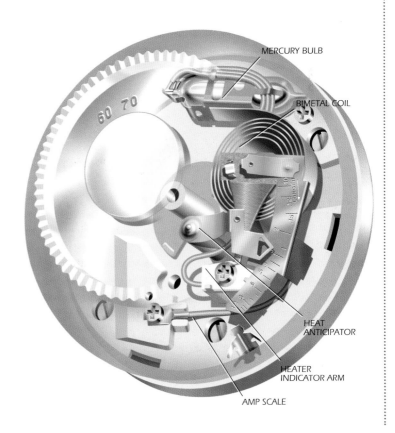

MERCURY BULB

BIMETAL COIL

HEAT ANTICIPATOR

HEATER INDICATOR ARM

AMP SCALE

The purpose of the heat anticipator is to prevent or reduce overshooting the temperature setting. Just before the temperature setting on a thermostat is reached, the thermostat's contacts open and shut down the burner. Heat remaining in the system brings the room temperature to the thermostat setting. The anticipator heats the thermostat's bimetal coil. In turn, the coil contracts (its diameter is reduced), and it moves the mercury bulb switch above it. Mercury in the bulb flows away from the electrical contacts inside it, opening the burner circuit.

Note in the drawing on the previous page that the anticipator's resistance is adjustable. By moving the heater indicator arm, you change the effective length of the anticipator, increasing or decreasing its resistance and the amount of heat it produces. If the anticipator's resistance setting is too high, too much heat is created. This will shut down the burner prematurely. If the resistance setting is too low, the opposite occurs.

The proper setting for the heat anticipator is one that matches the current draw of the thermostat control circuit. At this setting, the heat anticipator causes the thermostat to shut the burner at an ideal point.

The concept of the heat anticipator also applies to a hot-water or steam system, because with these systems, even when the burner is shut down, the residual heat in the radiators or convector units could cause the room temperature to overshoot the thermostat setting.

Noise on Ignition

I have been thinking about replacing my 27-year-old gas-fired heater. It works fine, but it makes a booming noise when it fires up. You know the old saying, "If it isn't broke, don't fix it." What do you think?

I don't think you should replace the heater, but you should check into the cause of the booming noise. One cause of this is delayed ignition. When an excessive amount of gas accumulates in the combustion chamber before ignition, the excessive fuel level causes a mini explosion.

There are a number of causes for delayed ignition that differ depending on whether you have an oil or gas-fired appliance. However, in both cases, the problem is usually with the fuel-delivery system as opposed to the heating appliance itself. In either case, the problem is hazardous and should be corrected by a licensed heating contractor.

Furnace Payback

Could you tell me how to calculate the payback on a gas furnace, based on the sticker efficiency rating and the cost of gas per therm? This would help me in comparison shopping for a new furnace.

ENERGYGUIDE—BRAND X

⬇

80,000 BTUH INPUT STANDARD FURNACE

CAPACITY: 58,000 BTU/HR

MODEL NO:

EFFICIENCY RATING

LEAST THIS MODEL   MOST
51.40   60.6%   81.19
▼   ▼   ▼

COMPARATIVE NATIONAL AVERAGE YEARLY COST ($) INFORMATION

COST PER THERM OF NAT. GAS	BTU/HR HEAT LOSS OF HOME (1,000s)						
	25	30	35	40	45	50	55
$ 10	87	108	120	137	154	170	187
$ 20	152	181	210	239	268	297	326
$ 30	217	258	300	341	383	424	465
$ 40	282	336	390	443	497	551	605
$ 50	347	413	480	546	612	678	774
$ 60	413	491	569	648	726	805	883

80,000 BTUH INPUT HIGH EFFICIENCY RECUPERATIVE FURNACE

CAPACITY: 69,000 BTU/HR

MODEL NO:

EFFICIENCY RATING

LEAST   THIS MODEL MOST
54.00   85.5% 86.20
▼   ▼   ▼

COMPARATIVE NATIONAL AVERAGE YEARLY COST ($) INFORMATION

COST PER THERM OF NAT. GAS	BTU/HR HEAT LOSS OF HOME (1,000s)						
	35	40	45	50	55	60	65	70
$ 10	90	103	116	129	141	154	167	180
$ 20	154	175	197	219	241	263	285	307
$ 30	217	248	279	310	311	372	403	434
$ 40	291	321	361	401	441	481	521	561
$ 50	344	393	442	491	540	589	638	687
$ 60	407	466	524	582	640	698	756	814

The most accurate way to calculate the payback between two new gas furnaces is, first, to do a heat balance on your house. That is, you should determine the heat loss per hour (in BTU/hr) in order to find the amount of heat necessary to maintain the desired temperature.

This analysis is quite difficult for the layman because the house's wall construction, floor square footage, exposure, window type and area, insulation, and so on must be considered.

A more practical way is to have a heating contractor give you an estimate of the heat loss. Also, you will need to call your local utility for the cost of gas per therm. A therm is a unit of heat equivalent to 100,000 BTU.

New gas heating appliances display a bright-yellow energy guide sticker. In the example shown, the heat-loss figures are the top horizontal line and the cost per therm of gas is the vertical left column.

For example, assume furnace X costs $1100 installed and the high-efficiency furnace costs $1500. If you pay 40 cents per therm, and the heat loss of your home is about 55,000 BTU/hr, the cost to heat your home for one year using furnace X is $605 and for the high-efficiency furnace, $441. The cost savings per year for using the more efficient furnace is $164. It would take 2.44 years for the fuel savings to pay for the cost difference between the furnaces.

Chimney Safety Tip

Prior to the heating season, it's important that you check your chimney to make sure that the carbon monoxide–laden exhaust gases from the heating system are not backing up into the house. Sadly, we occasionally read that a family has succumbed to carbon monoxide poisoning as a result of a clogged chimney. Carbon monoxide is colorless, odorless, and poisonous. The degree to which it poses a hazard is related to its concentration and the duration of exposure.

Use the following steps as a guide when checking your chimney at the beginning of each heating season.

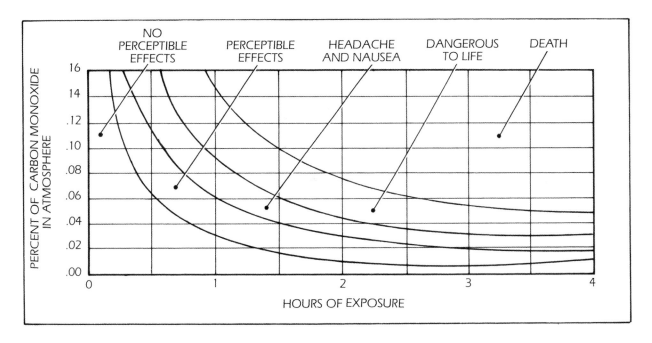

MISCELLANEOUS

1. Pull the vent pipe out and look into the opening. Remove any obstructions and clean away soot.

2. Hold a lit candle at, or blow smoke near, the chimney opening. If the flame or smoke is drawn into the chimney, the draft is okay. If not, the chimney is clogged.

3. Examine and clean the vent pipe that connects the furnace to the chimney. Make sure there are no breaks or holes through which exhaust can leak.

4. Properly replace the vent pipe so that it has an upward slope from furnace to chimney. Make sure that the pipe is not pushed in so far that it touches the back of the chimney.

Natural-Gas BTU

We buy natural gas by the cubic foot (and there are so many BTU per cubic foot). Our gas regulator is set at a pressure of 5 inches of water column. If the regulator were set at a higher pressure, would we be getting more heat per cubic foot of gas than at the lower pressure?

The heating value of natural gas is about 1000 BTU per cubic foot at atmospheric pressure. Increasing the gas pressure does increase the number of gas molecules per cubic foot. However, this is of no practical value. In fact, increasing the gas pressure to each appliance (water heater, furnace, boiler) serves no purpose because those appliances have a built-in gas pressure regulator. If the inlet gas pressure is too high, the regulator will reduce it to the pressure set by the manufacturer for optimum performance.

14. Domestic Water Heaters

SAFETY CONCERNS • RELIEF VALVE • WATER-HEATER ANODES • ELECTRIC WATER HEATERS • PREHEATER TANK • RECIRCULATING LOOP • TANKLESS COIL • MISCELLANEOUS

SAFETY CONCERNS

Water Heater Safety

I often do building inspections for prospective buyers. One problem I periodically run into is that of an improperly protected domestic water heater. I want to call this problem to your attention because it could result in an explosion causing severe damage or injury.

All tank-type water heaters, whether they are electrically heated, gas fired or oil fired, must have a relief valve that is both temperature and pressure sensitive. Unfortunately, many relief valves are pressure sensitive only, and this will not provide you with the needed protection against an explosion.

Should the heater's thermostat malfunction, the Water in the tank could become superheated, meaning it's heated beyond its boiling point of 212°F at atmospheric pressure. As the boiling point of water increases, its pressure increases and both combined can weaken a tank and cause it to rupture.

If this happens, the water pressure immediately drops to atmospheric pressure. The overheated water instantly flashes into steam, increases its volume, and liberates an amount of energy that can be greater than a pound of nitroglycerin. If the water had not been heated above 212°F, it would not flash into steam and an explosion would be impossible.

Consequently, pressure and temperature-sensitive

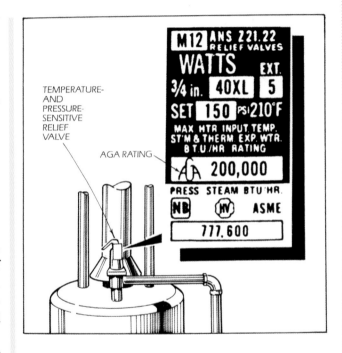

TEMPERATURE- AND PRESSURE- SENSITIVE RELIEF VALVE

AGA RATING

relief valves are designed to prevent the water temperature from exceeding 210°F. It does this by discharging the overheated water and allowing cooler water to enter the tank, safely lowering the water temperature.

It's also important that the Btu capacity of the relief valve exceed the Btu input of the water heater. If the capacity is less, the relief valve will not be able to discharge the overheated water at the same rate it's produced, resulting in an unsafe condition or possible explosion. Check the capacity of the relief valve by looking at the nameplate mounted on the valve. Specifically look for the AGA (American Gas Association) rating given in Btu/hour. This rating must exceed the Btu input to the water heater listed on the plate on the tank casing.

Also, when the relief valve is installed, its temperature sensing element should be immersed in the top 6 inches of water in the tank. This is important because there is a temperature difference between the tank and the hot-water outlet pipe. For instance, when the relief valve is installed in the hot-water outlet pipe 5 inches away from the tank, the water in the tank could be raised to 250°F before the temperature at the valve reaches 210°F.

Heater Safety

I have an electric water heater with a relief valve rated at 100,000 Btu/hour. The data plate on the tank casing contains the following wattage for the heating elements: Upper–4500, Lower–4500, Total–4500. However, there is no Btu input to the water heater listed on the plate. How can I tell if the relief valve is sized correctly for the water heater?

Electrical water heaters normally have an upper and a lower heating element. The data plate on the heaters will contain the maximum wattage of each element, as well as the total wattage of the unit. In most residential heaters, the heating elements do not operate simultaneously and, as a result, the maximum wattage of an element is the same as the total wattage.

To convert from electrical energy to heat energy, use this formula: 1 watt=3.413 Btu/hour. The total wattage of your heater is 4500. This is equal to an input rating of 4500 x 3.413 or 15,359 Btu/hour. Since the relief valve is rated at 100,000 Btu/hour, it's adequately sized for your water heater.

RELIEF VALVE

Dripping Relief Valve

I have a problem with my water heater. About 15 to 20 minutes after a large amount of hot water has been used, the pressure relief valve discharges water. I called a plumber, and he suggested installing an expansion tank in line with the water heater. In 30 years as a homeowner, having lived in eight different houses, I have never heard of this. Is the plumber's solution correct?

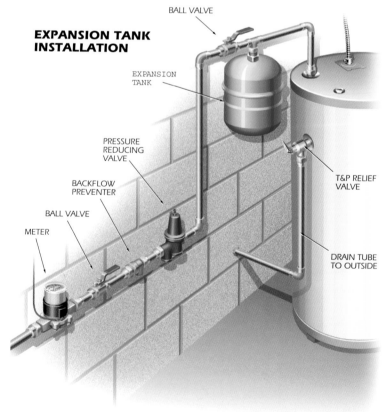

EXPANSION TANK INSTALLATION

BALL VALVE

EXPANSION TANK

PRESSURE REDUCING VALVE

BACKFLOW PREVENTER

BALL VALVE

METER

T&P RELIEF VALVE

DRAIN TUBE TO OUTSIDE

As water is heated, it expands. In this case, the water's expansion is probably being limited by a backflow-preventer valve or a non-bypass pressure regulator on the water inlet pipe. This results in a pressure increase that can be reduced one of two ways—by someone drawing hot water from a tap, or by the relief valve opening on the water heater. In your case, the relief valve discharges after the hot-water demand has passed because the water heater's thermostat is still causing the electrical heating elements to be activated or the burner to fire. The heater is producing hot water but the hot-water faucets are closed, creating a pressure buildup that is relieved by the valve. The relief valve discharges after a large amount of water is drawn because a small hot-water demand will not cause the heater's burner to fire or its element to be activated. Thermostats in residential water heaters let the water temperature drop 15° to 25° F before activating a burner or heating element.

Backflow preventers and non-bypass pressure regulators are valves that prevent contamination of a municipal water supply system. They stop water in the house's plumbing from flowing back into the municipal system.

I agree with your plumber. Installing a precharged, bladder-type expansion tank on the cold-water inlet between the pressure regulator and the water heater should correct the problem. The expanding hot water pressurizes the air in the tank, as opposed to increasing the system's water pressure.

Every house in my area has a pressure regulating valve (set for 60 psi) at the city water connection. When my water heater cycles on, the water pressure in the tank builds steadily and causes the water-heater relief valve to vent until a tap is opened, at which time pressure drops and the leaking stops. Is there a practical way to prevent the pressure buildup as described?

In order to prevent possible water contamination, many municipalities have code requirements for backflow preventer devices, including non-bypass pressure regulators on the inlet water service pipe. These devices create a closed system when a faucet is not drawing hot water. When water is heated, it expands. In a closed system where there is no room for this expansion, the pressure will increase. Consequently, the thermal expansion of water in the water heater will often result in dripping relief valves. Not only does this waste water, it also represents wasted energy.

In order to control this pressure increase, a provision must be made for the expansion. This can be done by installing a precharged bladder-type tank (expansion tank) similar to the ones used on hot water heating systems. The expansion tank should be installed on the cold water inlet, between the pressure regulator and the water heater. When the water expands, it flows into the bladder tank, thereby maintaining the pressure. When water is drawn from a fixture, the water from the tank flows back into the system. The tanks are generally available at plumbing-supply stores.

Relief-valve Piping

My home's water heater is in the basement just several feet from my library. If the heater's relief valve lets go, many books, along with furniture, would be damaged or destroyed by the hot water. A pipe runs from the relief valve down the outside of the heater and ends about 4 inches above the floor. There is a floor drain about 4 feet from the heater.

Is there a recommended (and easy) way to modify this system so the hot water could be contained, piped outside, or piped down the drain? Any suggestions for containing an overflow should one occur?

Discharge from the relief valve indicates a problem such as excessively high temperature, pressure inside the water heater, or even a faulty relief valve. Regardless of the cause, it's important that you know it exists, so you can correct it. If you piped the discharge directly into a drain or outside, you wouldn't know when a problem occurred.

When there is a problem, water will not flow continuously from the relief valve like a faucet. It flows in spurts, because as hot water is re-leased from the tank, cold water enters, lowering the temperature and pressure in the tank.

One solution is to put a pan under the discharge pipe. Normally, the pan will be dry' If you find water in it, you'll know there's a problem.

If you're concerned the pan will fill and overflow, you can run a drain line from the pan to the floor drain. Install the drain line about ¹/₂ inch above the pan's bottom. If you don't want the drain line to run on

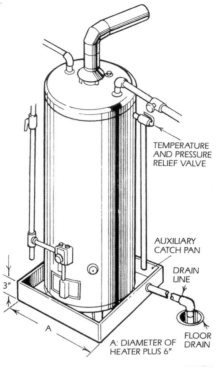

TEMPERATURE AND PRESSURE RELIEF VALVE

AUXILIARY CATCH PAN

DRAIN LINE

3"

A

A: DIAMETER OF HEATER PLUS 6"

FLOOR DRAIN

top of the floor, because it's a tripping hazard, and you don't want to chip out the floor to conceal it, then install a small pump in the pan, similar to an air-conditioner condensate pump. Run a discharge tube from the pump to any convenient drain or to the outside. Even after the pump has drained the pan, the residual water left will indicate there's a problem.

Of greater concern is whether an older water heater leaks. Water heaters generally last from seven to 10 years, but can last 20 years. If your water heater is that old, you might consider setting it on a collecting pan. This will solve the problem of containing water discharged from the relief valve. Inspect the pan periodically and drain it as discussed.

Municipal Water Pressure Prompts Valve Discharge

The relief valve on my water heater opens, and it got so bad that I had to use a 20-foot-long vinyl hose to discharge the water into the sump-pump basin. How can I prevent this waste of water and energy?

The problem you describe is a reasonably common one. Many municipalities require a backflow preventer on a building's inlet water service pipe to prevent contamination of the municipal water supply. When the faucets in a house are closed, the water distribution piping becomes a closed system. When water is heated it expands. In a closed system, there is no room for the expansion, so the water pressure will increase. This can cause the water heater's pressure relief valve to drip. To absorb this pressure, a pre-charged bladder-type expansion tank is installed on the cold-water inlet near the water heater. This is similar to the tanks used on hot-water heating systems.

When the water expands, it flows into the tank and presses against the bladder. The air behind the bladder is compressed, and the pressure in the house's water system does not increase appreciably. When water is drawn from a fixture, the water from the tank flows back into the system.

Relief-valve Location

In a previous column, you showed an indirect-fired water heater with the temperature-pressure relief valve (T&P valve) located on the hot-water outlet line. As I understand it, the valve should be installed in the top of the tank, and if conditions don't permit this, the valve should be installed on the cold-water supply and not the hot water outlet. If you disagree with this, please let me know why.

First, a word about water pressure and boiling-point temperature. Water at atmospheric pressure (14.7 pounds per square inch) boils at 212°F. As the water pressure increases, its boiling point rises. Water in a water heater is already at a pressure that exceeds 14.7psi. Typical residential water pressure ranges between 40 and 60 psi. As such, the boiling point of water in a water heater is higher than 212°F.

If the water heater's thermostat malfunctions, and causes an excessive heat buildup in the tank above 212°F the corresponding increase in water pressure can rupture the tank. If this happens, the water pressure would immediately drop to atmospheric pressure, flash into steam, increase its volume and liberate an amount of energy greater than an explosion caused by a pound of nitroglycerin. If the water had not been heated above 212°F, it would not flash into steam and an explosion would be impossible.

A T&P valve is a safety device that prevents the water temperature from exceeding 210°F. It does this by discharging overheated water through a pipe and allowing cooler water to enter the tank, thus safely lowering the water temperature.

**TEMPERATURE-AND-PRESSURE
RELIEF VALVE LOCATION**

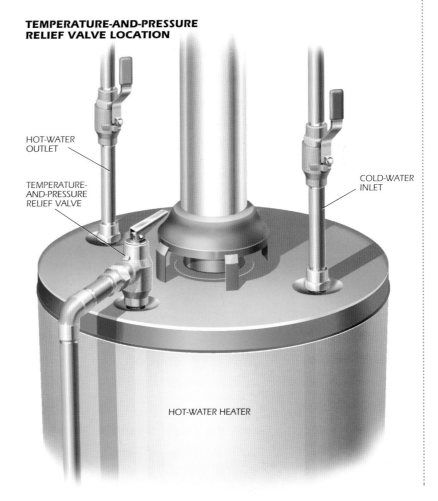

HOT-WATER
OUTLET

TEMPERATURE-
AND-PRESSURE
RELIEF VALVE

COLD-WATER
INLET

HOT-WATER HEATER

WATER-HEATER ANODES

I disagree with the installation of the valve in the cold-water supply line. In the event that the valve cannot be installed in or near the top of the hot-water tank, then it should be installed in the hot-water outlet pipe immediately above the tank. Because of a condition known as thermal lag, there is a temperature difference between the hot water at the top of the tank and hot water in the outlet pipe. For example, if the relief valve is installed in the hot-water pipe only 5 inches away from the tank, the water in the tank could rise to 250° F before the temperature at the valve reaches 210°F.

WATER-HEATER ANODES

The Function of Water-Heater Anodes

I cannot figure out why anodes are used in glass-lined water heaters. I have heard that the anode corrodes while the steel tank remains intact. But it is my understanding that no steel is exposed because the interior of the tank is covered with glass.

Water heaters, whether electric, gas, or oil fired have lined steel tanks. The lining is usually ceramic (glass) but in other cases it's concrete or copper. The tanks are lined to prevent the water from coming in contact with the steel and causing cathodic corrosion. However, because the lining may have imperfections such as an uneven distribution of the protective material or even tiny holes in it, the heaters are equipped with a sacrificial magnesium or aluminum anode that is suspended inside the tank to reduce tank corrosion. The electrochemical reaction that causes corrosion takes place

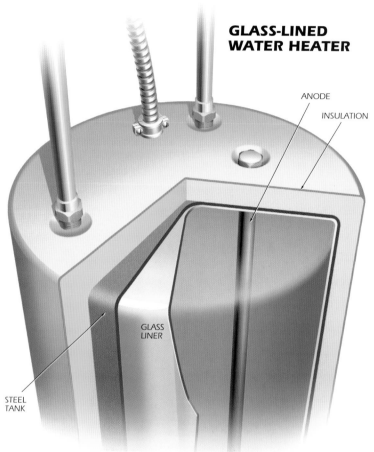

GLASS-LINED WATER HEATER

ANODE

INSULATION

GLASS LINER

STEEL TANK

between the water and the anode, rather than between the water and the steel tank.

Air in Hot Water

Every morning when I turn on the shower, there seems to be air coming out of the hot-water faucet. My plumber checked the water lines, pulled the well pump and inspected the water system's check valves. Everything seemed okay. Do you have any idea what causes this? The house is only a year old.

Since water-supply problems have been ruled out, let's take a look at the water chemistry and the water heater.

Your home's water chemistry may be such that it can react aggressively with metals, such as the anode in the water heater. This chemical reaction can create hydrogen gas. Only in extremely rare instances does this hydrogen gas pose a danger. It would take many months of the water heater being unused before the gas built up enough to reach dangerous levels. Other gases, such as oxygen, may occur naturally and be present in the water. When water is heated in a closed tank, its pressure increases.

As the pressure increases, the amount of gas that the water can hold in solution increases. When the hot-water faucet is opened, the water pressure decreases and the gas held in solution is liberated and forced out in spurts. Check with a plumbing contractor as to whether installing an expansion tank can correct this condition by reducing the pressure buildup in the system.

Hot-water Odor

I live in an apartment where there is an electric hot-water heater. When showering or washing at the kitchen sink, I detect a foul odor coming from the hot water. I can live with the smell, but I'd like to know what's causing it.

A domestic water heater basically consists of a lined steel tank. The lining is usually vitreous enamel (glass) but can be concrete (stone) or copper. Because the lining may have imperfections and pinholes, most heaters are equipped with a sacrificial magnesium anode rod that's suspended inside the tank to minimize tank corrosion. The electrochemical action that causes corrosion takes place between the water and the anode, rather than between the water and the tank. Therefore, the life of the tank is increased. Some tanks

are constructed so that the magnesium anode can be replaced if necessary.

The odor is probably the result of a reaction between the water and the magnesium anode. Water sometimes contains a high sulfate and/or mineral content. These chemicals can react with the anode and produce a hydrogen sulfide or rotten-egg odor in the heated water.

It's also possible that the odor is the result of the action on the anode of certain nonharmful bacteria in the water. In either case, chlorination of the water supply should eliminate or at least minimize the problem.

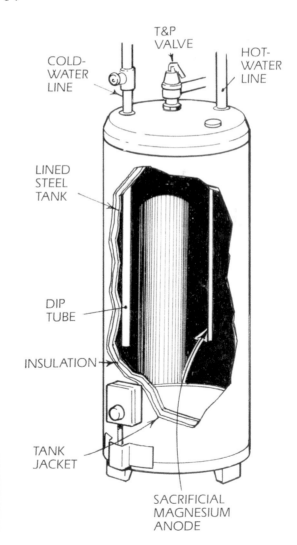

COLD-WATER LINE

T&P VALVE

HOT-WATER LINE

LINED STEEL TANK

DIP TUBE

INSULATION

TANK JACKET

SACRIFICIAL MAGNESIUM ANODE

Anode Replacement

I wanted to examine, and possibly change, the anode on my seven-year-old water heater. It wouldn't budge, at least not without rocking the heater. How can I get enough leverage on the plug to open it?

First, here's some background on water-heater anodes for readers who are unfamiliar with them. The anode is suspended in the water-heater tank to control corrosion. The anode corrodes, and in doing so prevents the water-heater tank from corroding. The anode consists of a magnesium or aluminum sheath extruded around a steel core called a stay rod. Although the anode starts out at ³/₄ inch in diameter, within three years it's reduced to the diameter of a pencil.

Installing a new anode is often difficult. Its hex-shaped top, sometimes called a spud, is often inaccessible or difficult to loosen. Usually you need a pipe installed over a wrench handle to provide enough leverage to free it. Also, depending on the heater's location and available headroom clearance, installing a rigid anode rod might mean disconnecting the heater from its plumbing and tipping it sideways. Flexible anode rods are available for these situations. Check with your local plumbing supply house.

In some cases, the anode is installed as part of the hot-water outlet. Replacing these anodes means removing the hot-water connection, but sometimes it's easier to add one of these outlet-mounted anodes than it is to replace the type shown here.

A new anode may extend the life of the tank, but you still have to determine whether or not this is cost effective. This depends on the age of the appliance and other factors. Keep in mind that water heaters have a variety of parts, aside from the anode, that wear out, and you have to judge whether it's better to prolong the life of the appliance or replace it. Most homeowners never replace the anode, and get 7 to 15 years of life from the appliance. However, some heaters last 20 to 25 years.

Finally, bear in mind that all water heaters—except those with artificially softened water—will get a slight lime-scale buildup inside the tank that provides some protection. Remember too that the life of your water heater also depends on the water quality and the temperature at which the water is maintained.

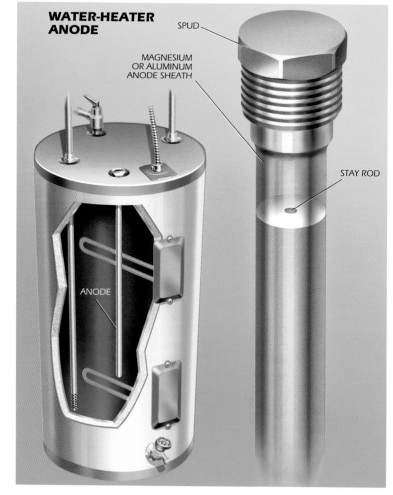

WATER-HEATER ANODE

SPUD

MAGNESIUM OR ALUMINUM ANODE SHEATH

STAY ROD

ANODE

ELECTRIC WATER HEATERS

Water-heater Timer

Installing a timer on your electric water heater can cut your utility bill by preventing the appliance from heating water when there is no demand for it.

To understand this, some background is helpful. Whether or not a water heater can supply an adequate amount of hot water will depend on both the capacity of the heater and the amount of hot water generated in one hour, known as the recovery rate. An electric water heater has a low recovery rate, generally on the order of 18 gallons per hour. As a result, in order to supply adequate hot water, the tank capacity is usually a minimum of 52 gallons and often can be as large as 80 gallons.

On the other hand, an oil-fired water heater can supply a comparable amount of hot water with a 30-gallon tank. These appliances have a high recovery rate, often on the order of 80 to 120 gallons per hour.

Homeowners who have electric water heaters can save energy and money by installing a timer that turns the water heater off during periods when hot water is not needed. The water heater will heat the water during the times of day when it is needed, such as for morning showers and evening dishes. The timer prevents energy from being wasted on storing a large quantity of hot water. Thus it helps the water heater match supply to demand.

A study conducted by the Illinois Institute of Technology showed that this approach can reduce the cost of heating water by 26 percent. Electric water heaters are among a household's biggest electricity users. In many cases, the timer could pay for itself in less than one year.

Slow Recovery

We remodeled our kitchen and replaced our 40-gallon, gas-fired water heater with an electric water heater that has a 5500-watt element. Now, we are stuck with a tank that is good only for me and my wife. When our three grandchildren come to visit, the water is not hot enough for them to bathe at half-hour intervals, like they did when I had a gas-fired water heater. Would you please help?

People switching from a gas-fired to an electric water heater frequently cite the problem you discuss. When selecting a water heater, consider its capacity and the recovery rate. The recovery rate is the number of gallons that the unit will heat to between 90° and 100°F above its inlet temperature in 1 hour. When the tank capacity is low, in order to have an adequate supply of hot water, the unit must have a high recovery rate. As an example, a typical oil-fired water heater has a 30-gallon capacity and a 120-gallon-per-hour recovery rate.

Electric water heaters, on the other hand, have a low recovery rate: usually about 18 gallons per hour, although some units have a 22-gallon recovery rate. With such a low recovery rate, unless there is a large tank capacity (on the order of 60 to 80 gallons), a simultaneous long-time demand for hot water will result in an inadequate amount of hot water being supplied to the fixture.

You can increase your heater's hot water output by installing a prewarming tank in series with the water heater. The cold-water supply is connected to the prewarming tank's inlet, and the tank's outlet is connected to the water heater's inlet. By boosting the temperature of the water entering the water heater, you improve its recovery rate. It is also more economical because you are not electrically heating a large volume of water all day, so there will be sufficient supply during the peak periods.

Erratic Water Heater

In order to save on utility bills, I set my new electric water heater at the energy conserving range of about 120°F. Each morning, the temperature of the water might be anywhere from 90° to 120°F. I've noticed that the temperature can drop to 85°F during the course of a five-minute shower. What's wrong with the water heater?

You may have a defective water heater. It's possible that the thermostat is out of calibration and needs to be replaced. However, if your unit has two heating elements, first check their settings. If the elements are not set at the same temperature, you can get variations similar to those you describe.

You didn't mention the capacity of the heater or its recovery rate. Electric heaters have a low recovery rate—especially in comparison to oil-and gas-fired units. If the tank has a low capacity, or hot water has been used prior to someone showering, then you may experience low hot-water temperatures.

Horizontal Water Heater: A Bad Idea

Can the average electric water heater be installed horizontally as a space-saving measure?

No. An electric water heater is designed for upright installation. Even shipping one in a horizontal position can damage it. The heater's anode and the water inlet pipe are not supported. Placing the water heater hori-

zontally can cause the pipe or the anode to flex against the tank's vitreous glass lining, damaging it.

Also, a horizontal installation would not be energy efficient. In an upright installation, the water inlet pipe terminates near the bottom of the tank. As the water is heated, it rises to the top of the tank where it is drawn off at the hot-water outlet. A horizontal installation would put the inlet and the outlet in the same plane and cause cold and hot water to be mixed.

PREHEATER TANK

Water Heaters Connected in Series

We have two 50-gallon electric water heaters connected in series, but we seldom use large amounts of hot water. If hot water is being dispensed one tank at a time, it seems that the water in the second heater in the series would rarely be dispensed. Could this standing water be a breeding ground for bacteria? How practical is it to turn off one via the breaker switch until large amounts of hot water are needed?

Yes, standing water can be a breeding ground for the bacteria that are always present in water. However, in a municipal supply, the bacteria are normally not toxic, but they can cause smelly or rusty water. If your water heaters are connected in series, you won't have standing water. As water is drawn from the first tank, makeup water will flow from the second tank into the first, and from the water supply into the second tank.

If you turn off the power to the second tank, that tank will function as a preheater by raising the water temperature to that approaching room temperature before it enters the first tank. This reduces your energy costs.

Readers who need more hot water should be advised that there are several ways to deal with the problem. For instance, a plumber may install the sec-

Preheating Water Heater

I have a 50-gallon electric water heater. However, I would like to switch to a 40-gallon gas-fired water heater. Would it be feasible (cost wise) to use my electric heater as a warming tank and heat the water by gas? How would I hook the two together as a unit? Any information you have on this subject would be appreciated.

If you have the space in your utility room for an extra tank, it would be worthwhile to use the electric water heater as a prewarming tank rather than discarding it, assuming the tank is sound and doesn't leak.

The tank should be connected in series with the gas-fired water heater. Connect the cold-water supply to the cold-water inlet pipe on top of the tank. The hot-water outlet pipe should then be connected to the inlet pipe on the gas-fired water heater. Also, disconnect the electrical wiring to the old water heater so that the unit functions purely as a storage tank. In order for the tank to be effective as a prewarming tank, you have to remove the insulation that surrounds the tank. This may be difficult because the insulation is located between the tank and the outer casing.

If condensation develops on the tank during the summer months, you can control it by installing a dehumidifier in the room. The advantage of a prewarming tank for the domestic hot water is twofold: It reduces the amount of gas needed to heat the water, and it increases the heater's recovery rate, which enables you to have more hot water over a given period of time.

ond water heater in parallel. Water heaters that are the same size and BTU input are usually installed this way-the advantage being that the system will provide large quantities of hot water at one time, since water is drawn from both tanks simultaneously. Also, this system allows you to shut off and drain one tank for periods where the extra hot water capacity is not needed.

Water heaters that are different sizes are often installed in series. The first and smaller tank functions as a preheater for the larger tank.

Other options for homeowners to consider would be installing the water heaters individually (so each serves a group of fixtures), installing an instantaneous water heater (a small gas or electric water heater adjacent to a tub/shower or sink) or installing a single large water heater. Each installation has advantages and drawbacks, and you should discuss each of these with the contractor who will perform the installation.

Water-tempering Tank

My electric water heater, although well insulated, is costly to operate. I have a wood stove that keeps the basement at about 70°F. Would locating a prewarming tank about 3¹/₂ feet away from the stove and tied to the water heater be a good idea?

An uninsulated water-tempering tank that acts as a reservoir for your water heater is a good idea—especially if the incoming water is very cold. Assuming that the prewarming tank has a 40-gallon capacity, then for every 1°F rise in water temperature that you achieve, you'll save approximately 100 watts. If the water entering the prewarming tank is 45°F and it's heated to 60°F before entering the water heater, then you'll save about 1.5 kilowatts.

Another advantage of a prewarming tank used in conjunction with an electric water heater is that it will increase the overall supply of available hot water. An electric hot-water heater typically has a low recovery rate in comparison to an oil- or gas-fired unit. The recovery rate is the number of gallons the unit will reheat in 1 hour. The recovery rate for electric hot-water heaters is about 18 gallons/hour (gph) whereas it's about 40 gph for a gas-fired unit and about 120 gph for an oil-fired unit. A higher recovery rate will enable you to have more hot water over a given period of time.

Of course, there is a trade-off with this system. You'll probably have to throw another log on the fire.

RECIRCULATING LOOP

Instant Hot Water

What can I do to provide instant hot water to every hot water faucet? My house is one level with the water heater at one end and the bathrooms at the other end, some 60 ft. away. The water pipes are attached to the floor joists.

If you want an instant hot water system, you'll have to install a recirculation line in the hot water piping. However, whether you can easily install a recirculating system depends on the accessibility of the pipes. Most homes have a noncirculating hot water system, which has the disadvantage of wasting time and water.

In your case, assuming there is a ¹/₂-inch-diameter water pipe between the water heater and the bathroom faucet, the 60 feet of pipe will contain about 1.35 gallons of water that are wasted while waiting for hot water (assuming you let this water run down the drain). The disadvantage of a recirculating hot water system is that it will result in a slight increase in your fuel bill. However, to minimize heat loss and conserve energy, the hot water distribution pipes can be insulated.

To convert your system to a recirculating type, you need to connect a ¹/₂-inch-diameter pipe to the hot

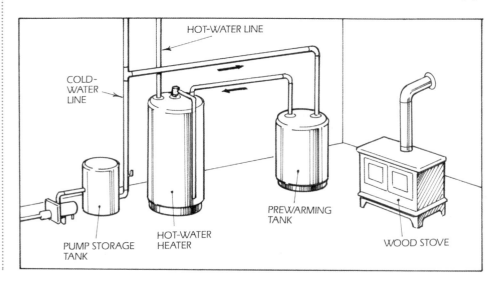

COLD-WATER LINE

HOT-WATER LINE

PUMP STORAGE TANK

HOT-WATER HEATER

PREWARMING TANK

WOOD STOVE

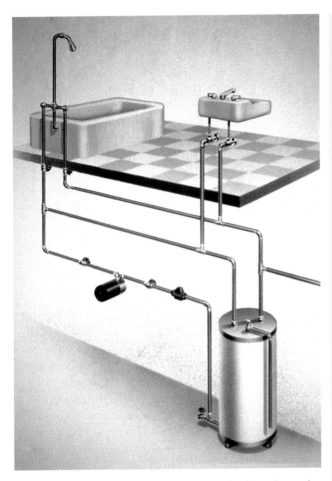

My problem is that I have a tankless coil in the boiler. Is it possible to install a recirculation line in this kind of a system?

With a tankless coil system, the return line should be connected to the cold-water pipe that goes into the tankless coil. Do not connect the recirculation line to the hot-water supply. That would lower the temperature of the hot water.

In past Homeowner's Clinics, you have explained a hot-water recirculation system. The only thing I didn't understand is where to make the connection for the recirculation line on the water heater. Is the connection made at the cold supply or the hot outlet?

The return line for a recirculation system should be connected to the cold-water supply line. If you connect it to the hot-water outlet, you will be mixing the hot water with cooler water. In a tank-type water heater, the cold-water inlet pipe terminates near the bottom of the tank. The drain value for removing sludge from the tank also is at the bottom of the tank. This is a convenient location to connect the recirculation loop. I've seen quite a few hot-water return pipes connected to a fitting on the water heater's drain valve.

water supply pipe at the fixture that's farthest from the water heater and run it back to the water heater. Several feet before the water heater, install a check valve, a globe valve, a recirculating pump and another globe valve. Next, remove the drain valve at the bottom of the heater and install a T fitting in the opening. Reconnect the valve to one end of the T and connect the return line to the other end.

The globe valves isolate the pump, should a repair or replacement of the pump be necessary. You can buy a recirculating pump at most plumbing/heating supply stores. The check valve is installed so the water flows in one direction—toward the heater. Otherwise, if the pump is not operating, comparatively cooler water from the bottom of the tank could enter the line, flow parallel to the hot-water supply line and out the tap at the sink or shower.

Hot-water Hassle

My water heater is at one end of my home and my bathroom is at the other end. In order to get hot water through the faucets, at least 80 feet of cold water has to come out of the hot water line. Is it possible to continue the hot water line past the bathroom and then return it into the water heater? Will this save a lot of water?

The hot-water distribution system that you have is a noncirculating type and is found in most homes. It is generally installed because it costs less for labor and materials than a circulating hot water system. Even though the noncirculating system is very common, it does have the disadvantage that you describe.

You can convert your system to a circulating hot water system by installing a return loop on the distribution line, which runs from the last faucet to the hot water heater. If the elevation difference between the hot water heater and the faucets is greater than 5 feet, then the hot water circulation can usually be achieved by gravity (this is because hot water rises, forcing cooler water down).

However, if it is less than 5 feet or if there is a long horizontal run, then you'll need a pump for effective recirculation of the water.

This system has the advantage of making hot water instantly available at all the fixtures because it is continuously circulating between the faucets and the hot water heater. This type of circulating system will also save some water.

Assuming there is a ¾-inch-diameter water pipe between the hot water heater and the bathroom faucet, the 80 feet of pipe will contain 1.8 gallons of water which would have been wasted. To minimize heat loss and conserve energy, don't forget to insulate the hot water distribution pipes.

Rusty Water

I installed a circulating pump in my hot-water system. Now the hot water is rusty. I have a timer on it so when we first turn on the hot water, it's very rusty. It never completely clears up. Before installing this circulation pump, I cleaned and flushed out the water heater, which is one year old. How do I get rid of the rust without having to buy a new water heater?

There are two possible sources for the rust, but the tank itself is usually not one of them. Most of today's water heaters have a high-quality glass lining, and it's very rare to have a lining failure significant enough to produce rusty water.

The rusty water is either caused by the circulator pump or iron reducing bacteria. If the pump has a cast-iron housing, rather than a brass or bronze housing, this could be the problem. If the problem is not the pump, then I would suspect iron reducing bacteria. Unlike other kinds of waterborne bacteria, this kind is not harmful. It is common in water-distribution piping in which soluble iron exceeds .2 ppm (parts per million). Soluble iron in the water provides food for the bacteria, and the discolored water is the end result of the bacteria feeding process.

The pump is probably stirring up the bacteria's waste products, which accumulate at the bottom of the hot water tank. Correcting the problem will require chlorinating the water heater and all of the hot-water system piping. If the water heater has a severe iron bacteria infestation, more than one treatment may be required.

TANKLESS COIL

Insufficient Hot Water

My hot water is heated by my boiler, which also provides heat for the house. However, the water is not even warm enough to take one shower. Do I need to replace the boiler or are there less expensive solutions?

There are a number of reasons why a heating system such as yours will not produce enough hot water and water that is not hot enough to be comfortable. Your system uses a tankless coil water heater. Here's how it works: A small water-filled coil is positioned in the boiler and surrounded by hot boiler water. The hot boiler water heats the water in the coil, and the water leaving the coil has its temperature increased by 90° to 100°F.

A tankless coil is designed so that the water will flow through it at a rate (in gallons per minute) that permits optimal heat transfer. If water flows too quickly through the coil, less heat will be transferred to each gallon. This results in a less-than-comfortable water temperature. This can be corrected by installing a flow regulating value on the cold-water supply pipe to the coil.

Also, many hot-water systems have a mixing valve. As its name implies, this device mixes the hot water leaving the coil, with cold water and ensures that the water is at a safe temperature before it travels to fixtures throughout the house. The mixing valve can be adjusted to increase the water's temperature.

Mineral formation may also cause low-temperature water. Over the years, mineral deposits tend to build up inside the tankless coil, especially if the water has a high mineral content. The deposits act as an insulator and prevent proper heating of the water in the coil.

Finally, if you have a steam boiler rather than a hot-water boiler, it's possible that the water in the boiler is not high enough to cover the tankless coil.

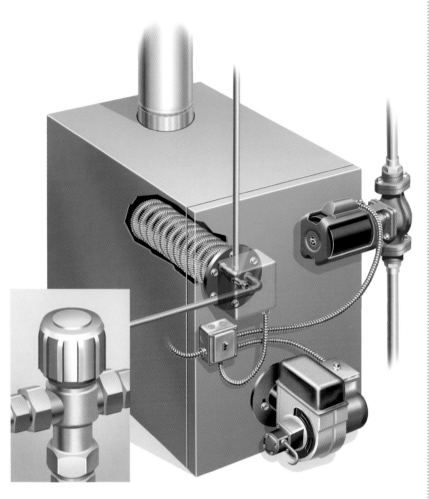

My house's hot water is supplied by a tankless coil connected to a boiler. The system supplies an inadequate amount of hot water. I am about to replace my heating system, and I want to know how to correct this.

If you are going to replace the boiler, I would recommend that you do not get one that produces heat for the house and hot water. My preference is a hot-water boiler for comfort heating and a separate water heater that is fired by oil or gas.

My reasoning is as follows. Most tankless water heaters consist of a small-diameter coil of pipe inside the boiler. The coil is surrounded by hot boiler water that gives up heat to the water flowing through the coil. Cold water enters the coil and leaves with its temperature increased by about 100°F. The system is designed with a flow rate through the coil that achieves the desired temperature.

However, if water flows too quickly through the coil, less heat will be transferred to the water in the coil. Thus, many tankless coils have a flow-regulating valve installed in their cold-water supply pipe. In addition, over the years mineral deposits tend to build up inside the heater coil, especially in hard-water areas. This reduces the water flow through the coil.

Because a tankless coil has virtually no storage capacity, the result is that household demand frequently exceeds hot-water supply.

Hot-water Supply Problem

If I turn on the hot water in my house when someone is taking a shower, they'll get an unexpected cold blast. Can this problem be explained by the small-diameter ½-inch pipe in the heating loop? Or, is it caused by lime encrustation in the heater? The cold water supply pipe is ¾ inch diameter.

It sounds like your hot water is generated by a heating system that uses a tankless coil rather than an oil- or gas-fired tank-type water heater. The problem you describe is common in units where there's a mineral buildup on the inside of the coil. The lime encrustation greatly reduces the effective opening and, even when the pressure is good, reduces the water flow.

The minerals can usually be removed by an acid flush. This, however is not recommended. The coil may be pitted, and removing the encrustation could cause it to leak. Your best bet is to replace the tankless coil. Also, the water flow in the ½-in. pipe is slightly less than half the flow in your ¾-inch supply pipe. The ½-inch pipe was probably installed to reduce the flow so that the water can reach the desired temperature. If cold water flows through the coil too rapidly, it not absorb the required heat.

If, after replacing the tankless coil, the hot-water flow is less than you desire, I suggest you replace your hot-water system with an oil- or gas-fired, tank-type unit and replace the ½-inch pipe with ¾-inch pipe.

Auxiliary Hot-water Tank

Our domestic hot water is provided by a coil in the boiler, a common method which is inadequate when a large quantity of hot water is wanted, since there is no storage beyond the small contents of the coil. Is there a way that a well insulated storage tank could be connected into the system and kept hot by thermal convection from the present furnace coil? Would any special precautions or valving be required?

Yes to both questions. A tank will greatly improve the effectiveness of your hot-water system. The tank, which should have a minimum capacity of 30 gallons, can be connected directly to the hot-water coils. If possible, the tank should be positioned above the coil outlet so that the hot water will circulate between the tank and the coil by gravity (hot water rises, forcing cooler water down).

If, because of space limitations, the tank cannot be positioned above the coil, recirculate the hot water by using a circulating pump. There is one big precaution: There must be a relief valve either on the tank or on the hot water outlet pipe which is both temperature and pressure sensitive.

Fluctuating Water Temperature

Our 10-year-old boiler was set up by the house's previous owner so that domestic hot water is preheated first in an electric water heater. We've noticed that the water starts out very hot, then cools off rapidly until it is uncomfortable. Two people cannot take a shower one after the other. This seems to be an inefficient arrangement. Is there a way to improve it?

When the hot water is generated through the heating system, the boiler operates in the summer and winter. The previous owner probably used the electric water heater by itself in the summer to produce hot water. In the winter, he may have continued to use the water heater as a preheater, but with the power shut off. Preheating the water in a storage tank so that the water approaches the basement air temperature is considered energy-efficient. However, preheating the water electrically is not efficient, and it is usually not necessary.

Fluctuation in water temperature and a shortage of hot water is a problem associated with older boilers that produce the hot water in a coil inside the boiler. In these cases, water enters the coil and leaves with its temperature raised by 90° to 100°F. The system is designed with a specific water-flow rate through the coil, and if water flows through the coil at a faster rate, less heat is transferred to it. Consequently, many boilers have a flow-regulating valve that restricts water flow.

Normally, when hot water is generated in a coil, the problem is not enough hot water for simultaneous demands. It shouldn't be a problem when the demand is sequential, such as one shower after another. Your problem may be caused by cold water flowing through the coil too rapidly to absorb the heat. It is also possible that lime deposits have built up in the coil. The deposits act as insulators and reduce heat transfer.

Consult a plumbing and heating contractor about the best way to solve the problem. Increased efficiency in water heating is not only more comfortable, but often pays for itself in reduced utility bills.

MISCELLANEOUS

Hot-water Temperatures

Most hot-water tanks have a dial for water temperatures at their bottoms. They read HOT, WARM, and NORMAL. What are the temperatures at the three settings? I've heard of a code in most places that says the minimum temperature should be 120°F and the maximum 140°F.

Not all water-heater manufacturers use the same names for the thermostat settings. Nevertheless, the settings are basically HOT, MEDIUM, and WARM. On some water heaters there is also a vacation setting. According to the manufacturers, the respective temperatures are 160°, 140°, 120°, and 60°F. These temperatures are not precise, as there can be a difference of 10°F or more between the dial setting and the water temperature.

The thermostat for a tank-type water heater is at the lower portion of the tank. Since hot water rises, the temperature of the water at the top of the tank (where the hot-water outlet pipe is located) will be higher than the water surrounding the thermostat. Also consider that hot water loses heat as it flows through the distribution pipes.

For the most part, a thermostat setting that produces a water temperature of 140°F will be adequate for household appliances such as clothes washers and dishwashers. However, when dishes are to be washed by hand or when bathing, a water temperature of 120°F is probably too hot for most people and needs to be tempered with a bit of cold water. (If you have small children in your house, consider keeping the maximum temperature at 120°F.)

A thermostat setting higher than 140°F wastes energy and shortens the water heater's life. A water temperature in excess of 160°F is a potential hazard because a person can be scalded while showering if the shower's mixing valve is faulty.

Rumbling Water Heater

I have a gas-fired, tank-type water heater. Lately, it has been making a rumbling noise when it fires. That's the only time it makes this noise. Should it be replaced?

The water heater does not have to be replaced. Over the years, sediment, scale and mineral deposits accumulate on the bottom of the heater tank. Manufacturers suggest that a few quarts of water be periodically drained from the heater to help remove these deposits. However, this practice is not always effective, and, if sufficient deposits accumulate, a rumbling sound can be heard when the unit is firing. It is not a dangerous condition, although it can be annoying. Also, the deposits act as an insulator between the water and the flame, and decrease the heater's efficiency.

Noisy Water Heater

Our eight-year-old propane-gas-fired water heater has been making crackling and bumping noises in recent months. The noises start when hot water is drawn and continue for a time after the flow stops. Any help you can lend is greatly appreciated.

Your appliance has a condition that is common among old water heaters: the formation of sediment and scale on the base of the tank. Manufacturers suggest that a few quarts of water be drained from the heater periodically to help remove the deposits. However, this is not always effective, and if sufficient deposits accumulate, a rumbling or pounding sound can be heard when the heater is firing. It's not dangerous, but can be annoying. Also, the deposits reduce the heater's efficiency because they insulate the water in the bottom of the tank from the flame.

Rotten-egg Odor

Even though we clean our bathroom twice a week, it has a rotten-egg odor. Where is it coming from?

The odor has nothing to do with cleanliness. It's probably hydrogen sulfide gas, which results from high concentrations of sulfates in the water. Sulfate-reducing bacteria convert the sulfates to sulfides. The bacteria, which are not harmful to humans, flourish in the warm water of the house's water heater tank. Chlorinating the water heater using a chlorine feeder usually eliminates the bacteria. You should check with a plumbing supply house for this equipment.

When to Drain a Hot-water Heater?

I live in Florida six months of the year. When I return to my home in Florida, the hot water smells very bad, and I have to drain the water heater to remove the smell. Should I drain the hot water tank prior to closing the Florida house?

The best solution is to drain a hot water tank prior to leaving a house vacant for six months. As a means to prevent the problem you discuss, some water quality experts and water heater manufacturers recommend that a tank be drained if it goes unused for two months.

The most common cause of "smelly-water is non-toxic, sulfate reducing bacteria that thrive in water under two specific conditions: a temperature below 138°F and the presence of hydrogen gas (given off by the water heater's anode as it is depleted).

Water-heater Insulation

For several years, utility companies have recommended installing insulation blankets on our electric water heaters. Some even do it free for customers. I installed one on my own, but I was recently told that the blanket will drastically reduce the life of my heater by trapping condensation and causing it to rust. I was advised to remove the blanket and use the escaping heat to prevent my laundry room pipes from freezing. Whom do I believe?

All tank-type water heaters are insulated by the manufacturer. The insulation is between the storage tank and the outer casing. Even with this insulation, the surface of the outer casing becomes warm to the touch, and as a result it gives up some heat to the room in which it's located. You can reduce this heat loss by covering the casing with an insulated blanket, as was recommended by your utility company.

Condensation is not a problem you have to worry about. For the water vapor in the air to condense, it must contact a surface that's cooler than the room's temperature. Since the temperature of the water heater casing will be either the same as room temperature or above, moisture in the air will never condense on it, regardless of whether or not the casing is insulated.

If you are concerned about pipes freezing in the laundry room, install a thermostatically controlled space heater, rather than depending on the water heater. This way you can regulate the room temperature to suit your needs and the time of year.

Slimy Hot Water

Our hot water storage tank recently leaked and was replaced by another stone-lined tank. My family and I noticed an unsettling change in the feel and texture of the water in the shower and sinks the evening of the installation. My wife described the water as feeling slimy or soft. I thought it felt more silky, such as diluted mineral oil or water that had portland cement residue mixed in with it. Nevertheless, the water was nice and hot. What can you tell us about stone-lined hot water storage tanks? Will the residue rinse away with time, and is it a health hazard?

The term stone-lined is really a misnomer. The storage tank is actually lined with concrete. Probably, the company that manufactured your unit uses a fine sand in the concrete mix, which has the texture of flour.

You were on the right track when you said the water had a cement residue feel to it. Any tank residue should flush out shortly. If it doesn't, contact the manufacturer.

There is nothing inherently unsafe with stone-lined hot water tanks, but, if you are concerned, you can have the water tested to see if its mineral content exceeds the EPA standard for potable water.

Water-heater Venting

I just replaced my gas hot-water tank. The instructions for the new tank recommend that at least once a year I inspect the venting system to look for obstructions, damage, and rust. I checked the visible portions of the vent, but most of it runs through interior walls. Is there an alternate way to check the venting system other than visual inspection?

The vent system for a water heater should discharge exhaust gas harmlessly outside, and not "spill" it back into the utility room. In addition to a visual inspection, you should check the area by the draft diverter (at the base of the vent pipe) for escaping exhaust gas. Put your hand close to the opening at the top of the water heater. If exhaust gas spills out of the diverter, you will feel hot gases blowing across your hand. This is a condition that must be corrected because the gas contains carbon monoxide. This colorless, odorless gas is poisonous and can cause asphyxiation.

Another check for faulty venting is to hold a lit match at the draft diverter opening. If the exhaust gases escape into the room, they will blow out the flame. If the vent system is functioning properly, the flame will be sucked into the draft diverter hood.

As a safety precaution, I recommend that homes with fuel-burning appliances be equipped with one or more UL-listed carbon monoxide detectors. They can be purchased at home centers and hardware stores.

Venting Gas Appliances

In a recent column you commented on the necessity of venting natural-gas water heaters. Why do some gas-fired appliances need to be vented and others don't? I have seen wall heaters that are not vented outside, for example.

Generally speaking, a gas appliance is vented to the outside if it produces enough carbon monoxide to pose a threat to occupants of the dwelling. Kitchen ranges, for example, are not usually vented outside because they tend to be located in well-ventilated areas. Also, they tend not to be left burning for long periods. However, there are times when even these appliances can pose a hazard, such as when people keep them burning day and night during a power outage.

Although some gas heaters are vent-free (they do not vent outside), most do require venting. It's not always easy to tell whether an appliance is vented. Some water heaters and even furnaces are referred to as direct vent. These vent directly to the outside through a short section of metal or plastic pipe.

Solar-heated Hot Water

I live in a warm climate. The temperature never drops below 40°F. I'd like to install two hot-water-heating solar panels on my roof and connect them to my electric hot-water tank so that hot water from the roof enters the tank through the cold-water inlet. Could you please discuss this in your column?

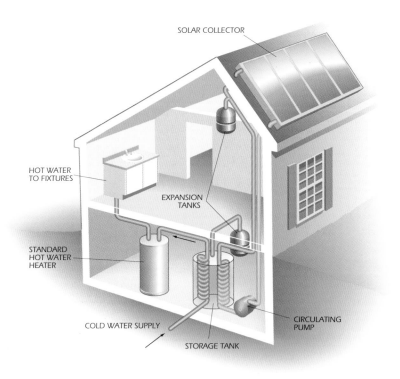

SOLAR COLLECTOR

HOT WATER TO FIXTURES

EXPANSION TANKS

STANDARD HOT WATER HEATER

COLD WATER SUPPLY

STORAGE TANK

CIRCULATING PUMP

Over the years, a number of designs have evolved for heating water with solar panels. A generalized diagram for one such system, which can be used in locations that have below-freezing winter temperatures (most of the country) is shown. This system has an antifreeze solution circulating through the solar panel. Note that the storage tank is a heat exchanger—it transfers heat from the antifreeze solution to the domestic water. By using two sets of coils, the system prevents the potable water supply from being contaminated if antifreeze should leak out.

In areas where the temperature never drops below freezing, you can run potable water from the collector directly into the water heater. Also, some designs do away with a water heater. The collector's capacity is large enough so that it serves directly as the water heater.

Collectors are usually installed facing south and tilted at an angle. Since the angle of the sun changes with the seasons, the collector's angle becomes a compromise. A rule of thumb is that the collector panels should have a tilt equal to latitude plus 5 degrees, measured from horizontal.

If the collector panels are placed on the roof, make sure to mount them on brackets that will provide adequate strength. Solar panels are subjected to uplifting forces when wind strikes their undersides, and should be built so that they can withstand gusts of up to 100 mph.

CENTRAL AIR

Condensate Leakage

I have an air-conditioning unit in my attic. Although the unit has a drainpipe, condensate from the air conditioner is leaking and wetting the ceiling of the room below. Do you have any idea what causes this?

Most air conditioning systems in homes are split systems. That is, the compressor/condenser is located outside the house and the evaporator's cooling coil and blower are located inside—in the attic in many cases. The evaporator should be mounted on a vibration-dampening pad and positioned over a pan. The pan is used as a backup for condensate collection. In the event that the condensate line becomes clogged, the condensate will leak out of the air conditioner and be collected in the pan. Check your unit. The lack of a pan could be the source of your problem.

Also, there should be two independent drain lines: a primary condensate drain and a backup or auxiliary drain. A common, but incorrect, installation practice is to connect the two drain lines. In this case, if the primary drain line is clogged, the backup or auxiliary drain cannot perform its intended function.

A malfunctioning switch is another possible source for leakage. In some installations, instead of a drain line there is a micro-switch installed in the pan. As the water level in the pan rises, it trips the switch, which

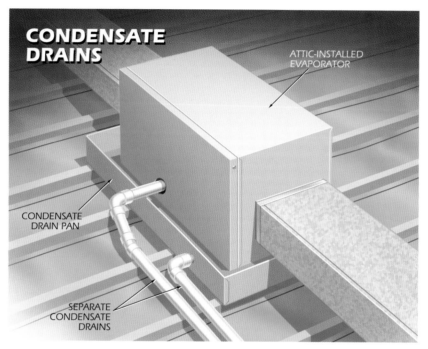

CONDENSATE DRAINS

ATTIC-INSTALLED EVAPORATOR

CONDENSATE DRAIN PAN

SEPARATE CONDENSATE DRAINS

shuts off the power to the air conditioner. This prevents an overflow condition. However, I've been told by some air conditioning servicemen that the micro-switch occasionally malfunctions.

One-story Central Air

We live in a 1200-square-foot, one-story house, and plan to add a family room next to the kitchen. We have an oil-fired, hot-water heating system with baseboard radiators. Would it be practical to centrally air-condition our house? Can the ducts be run through the attic? And how would it be powered?

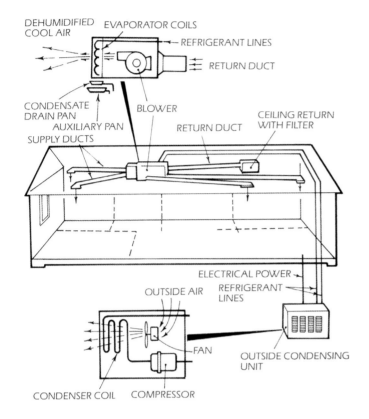

DEHUMIDIFIED COOL AIR
EVAPORATOR COILS
REFRIGERANT LINES
RETURN DUCT
CONDENSATE DRAIN PAN
AUXILIARY PAN
SUPPLY DUCTS
BLOWER
RETURN DUCT
CEILING RETURN WITH FILTER
ELECTRICAL POWER
OUTSIDE AIR
REFRIGERANT LINES
FAN
OUTSIDE CONDENSING UNIT
CONDENSER COIL
COMPRESSOR

Duct Location

I want to install central air conditioning in my home. Since I have hot-water heat, I will have to install air ducts. Is it more efficient to put the ducts in the ceiling or in the floor? Would I lose cooling efficiency if the ducts are installed in the attic? It gets extremely hot up there.

Don't worry about the loss of cooling efficiency from the warm attic. As long as the attic is adequately ventilated and the ducts are insulated, the heat that the ducts gain will be minimal. If you want to reduce the heat load in the attic, install a thermostatically controlled power ventilator in the roof.

The best place for the supply registers is on the ceiling or high on the wall. Cool air leaving the register will sink and displace warm air, which rises. The return grille should also be in the ceiling or high on the wall. By installing the supply registers and return grilles in this way, the air currents will mix. This prevents layers of warm and cool air from forming.

In many cases, central air conditioning is installed using heating ducts to move cool air. Although this saves the cost of installing a set of air conditioning ducts, it is not as efficient. The cool air entering through registers at floor level forms a cool layer with warm air above it.

A better installation, though more expensive, is to have ducts installed in the walls with high/low registers. In this system, there is one register slightly above floor level and another high on the wall. Heated air is discharged near the floor, while cool air is discharged near the ceiling. Airflow is directed to the appropriate register by moving a damper that is part of the high/low system.

It is possible—in fact, relatively easy—to install central air conditioning in a one-story house. The basic components of an air-conditioning system are the compressor/condenser, ducts, and a blower coil, or air handler.

The compressor/condenser is usually located outside the house. Refrigerant lines run from the compressor/condenser to the blower coil—up the outside wall and along the attic floor.

Unless the attic is extremely low and inaccessible, the blower coil is usually located there. The ducts run along the ceiling beams to ceiling registers located in the various rooms.

When a blower coil is located in the attic, mount it on an insulating material to minimize vibration noise. It should have an auxiliary condensate drain pan located below the unit to collect overflow if the primary condensate drain becomes clogged. If your electrical service to the house is not at least 100 amps at 220 volts, it will have to be upgraded.

Condensate Drainage

My basement condensate pump evacuates the condensate from my central air conditioner. But the pump is fairly noisy and can be heard in the living area. Can I simply bypass the pump and run the drain hose into a hole bored through the basement slab? The basement is bone-dry. There are 12 inches of crushed stone under the slab and footing drains run into a sump pit. Is there any reason why I shouldn't do this?

Generally speaking, it's always preferable to pump condensate from a basement. I would be concerned about your solution if there were problems with a high water table. However, since your basement is dry, your solution should work. Frankly, though, I'm surprised that the pump is so noisy. It's possible that it's defective. Check into this before proceeding.

If you go ahead with the solution that you propose, be sure you're not violating local building codes, and don't make the drain hole any larger than it needs to be.

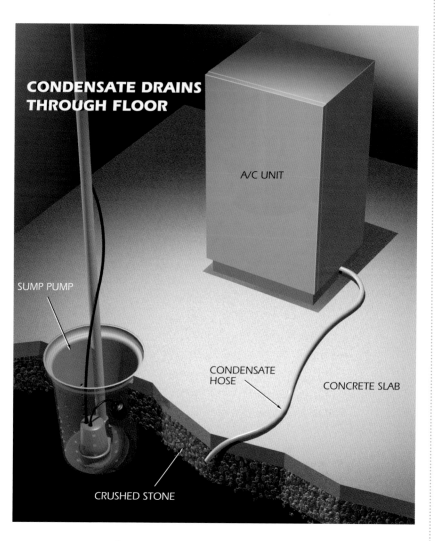

CONDENSATE DRAINS THROUGH FLOOR

A/C UNIT

SUMP PUMP

CONDENSATE HOSE

CONCRETE SLAB

CRUSHED STONE

Air-conditioning Humidistat

What is your opinion on installing a humidistat in series with my air-conditioner thermostat? Will this save on operating costs?

The amount of electricity saved will vary considerably with the user. There is a disadvantage in connecting the controls in series. In this case, both the humidistat and the thermostat contacts must be closed for the air conditioner to operate. If the relative humidity is low, the air conditioner will not run even though the room may be uncomfortably hot. When the two controls are connected in parallel, the air conditioner will operate if the contacts on the humidistat or thermostat are closed. The condition that is met first, temperature or humidity, triggers the air conditioner. Here, you can use the air conditioner to control humidity by setting the thermostat to engage at a high temperature.

WALL AND WINDOW UNITS

Air-conditioner Sizing

Is there a rule of thumb for accurately sizing air conditioners?

There is no simple rule for calculating the required size of a room air conditioner—at least, one that is accurate. There are many factors that affect cooling requirements, such as the number and size of windows and whether they are single, double or triple glazed. There are sizing charts available that can be used as guidelines. The following figures have been developed by the New York State Energy and Development Authority and the federal Energy Star program.

Square Feet	Btu/Hr.
100 to 150	5000
150 to 250	6000
250 to 300	7000
300 to 350	8000
350 to 400	9000
400 to 450	10,000
450 to 550	12,000
550 to 700	14,000
700 to 1000	18,000

If the room is naturally cool, reduce the Btu/Hour figures by 10 percent. Conversely, if the room is naturally warm (if it has a southern exposure, for example), increase capacity by 10 percent. If more than two people regularly use the room, add 600 Btu per hour for each additional person.

An air-conditioning system should be properly sized or slightly undersize, but it should not be oversize. When an air conditioner is too large, it operates intermittently, quickly chills the air, and then shuts down. When the system is shut off, moisture in the air does not condense, and the system does not dehumidify the air.

16. Environmental Concerns

RADON • MOLD

RADON

Radon Ventilation

I am concerned about radon entering my house through the basement. What can a homeowner do to reduce levels? And what levels are considered high?

Radon is a colorless and odorless radioactive gas given off by decaying uranium deposits within the earth. The gas is measured in picocuries per liter of air. Federal environmental officials recommend taking measures to reduce the concentration of radon when it reaches 4 picocuries per liter (4 pCi/L).

To do this, have a contractor who is state-certified or listed with the federal Environmental Protection Agency's radon-measurement proficiency program determine the year-round average radon level in your house. This is done using a long-term test of 90 days or more, which will produce more reliable results than a short-term test of two days to 90 days.

If a long-term test yields radon concentrations greater than 4 pCi/L, then the agency recommends corrective action. One very effective method is to seal cracks in the foundation wall and basement floor, and remove the radon with a ventilation system such as the one shown. Holes are bored in the house's basement slab, pipes are inserted in the holes and sealed to the slab. An inline fan pulls the gas from below the slab and exhausts it out a roof vent.

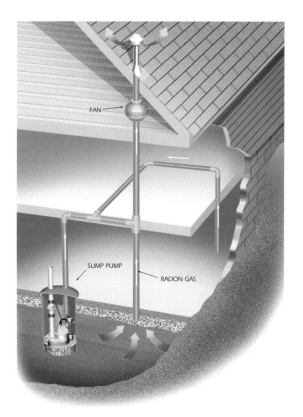

FAN

SUMP PUMP

RADON GAS

Radon Test Results

How serious a problem is radon? The short-term test reading in my house was high. Also, what's the best way to fix the problem?

For readers who are familiar only with the generalities regarding radon, perhaps some background is useful. Radon is a colorless, odorless gas that occurs from the decay of radium, a metallic, radioactive element. The degree of threat that it poses to your health is a matter of some controversy. Its potential as a health threat was extrapolated from the health statistics of uranium miners, specifically the rate of lung cancer among them. Many scientists believe this lung cancer data can be used to predict the risk that radon gas poses to the general population. Other scientists feel that the uranium miner data is not useful as a statistical tool. However, attempts have been made to show that regions with high radon gas concentrations have higher-than-normal concentrations of lung cancer in the general population.

We're not in a position here to tell you whether radon is a health threat. All we can say is that the Environmental Protection Agency considers it to be one. As a result, selling your house could be difficult unless you install a radon-removal system. Fortunately, this is relatively easy to do. Before we get into that discussion, you may find the following information helpful.

It's standard practice today to take radon mitigation measures when a short-term test measurement is higher than 20 picocuries per liter of air (pCi/L). An effective method of reducing that level is to install a system that uses fans and piping to remove the radon from beneath a house's basement slab.

In cases in which the radon reading is less than 20 pCi/L, a long-term test is recommended to determine the radon level more accurately. A long-term test spans one heating season and lasts anywhere from 3 to 12 months. If the long term test shows a level of 4 pCi/L or higher, it's common practice to install a radon ventilation system such as the one discussed above.

MOLD

Questions about Mold

I have heard many reports of mold in houses in the Houston area. What is mold and how can I prevent it from occurring in my house?

Mold is the common name for simple plants or fungi that are usually microscopically small. There are many varieties—some estimates put the number at a mini-

MOLD SOURCES

STACHYBOTRYS

PLUMBING LEAK

ROOF LEAK

SURFACE DAMPNESS

TRAPPED MOISTURE VAPOR

WATER SEEPAGE

mum of 40,000. These plants thrive on cotton, wool, paper, leather and wood, and they will grow anywhere in the home where there is moisture. The moisture may be in the material itself or in the form of airborne vapor. A typical mold is the one found in the grouted tile joints of shower walls.

Most mold spores found in homes do not present a serious health problem. However, people who have mold allergies could experience symptoms such as nasal stuffiness, eye irritation and difficulty breathing.

In the mold-related incidents you are referring to, severe moisture problems in a number of houses led to a comparatively rare mold condition—that is, a concentration of the mold species called *Stachybotrys chartarum*. The family living in one home suffered a number of severe health-related problems.

There isn't room here to present a complete discussion of mold, but the drawing above gives an overview of the parts of your house that require careful inspection and maintenance. Pay particular attention to the roof and to chimney flashing to be sure that neither is admitting water. Likewise, leaks around doors and windows can admit water as can cracks in a basement floor. Condensation on windows, leaking plumbing and moisture vapor from a crawlspace are other sources that can cause mold to form. The best way to control mold is to keep building surfaces and cavities as dry as possible.

GLOSSARY

Backfill The gravel or earth replaced in the space around a building wall after the foundation is in place.

Backflow The flow of a liquid (water) in a direction opposite to the natural or intended direction of flow.

Bottom plate The bottom horizontal member of a frame wall.

Btu (British thermal unit) The amount of heat required to raise the temperature of 1 pound of water by 1 degree Fahrenheit.

Control joint A groove that is formed, sawed, or tooled in a concrete or masonry structure to regulate the location and amount of cracking.

Cricket Also called "saddle." A small saddle-shaped projection on a sloping roof, used to divert water around a chimney.

Cross connection A direct arrangement of a piping line that allows the potable water supply to be connected to a line that contains a contaminant.

Cupping An inward-curling distortion at the exposed corners of asphalt shingles.

Downdraft A downward current of air in a chimney, often carrying smoke with it.

Dry well A covered pit, either with open-jointed lining or filled with coarse aggregate, through which drainage from downspouts or foundation footing drains may seep into the surrounding soil.

Eaves The lower edge of a roof that projects beyond the building wall.

Efflorescence A white powdery substance appearing on masonry wall surfaces, composed of soluble salts that have been brought to the surface by water or moisture movement.

Fascia (or facia) A horizontal board nailed vertically to the ends of roof rafters; sometimes supports a gutter.

Flashing Sheetmetal or other thin, impervious material used around roof and wall junctions to protect joints from water penetration.

Flue A passageway in a chimney for conveying smoke, gases, or fumes to the outside air.

Frost line The depth of frost penetration in soil; varies in different parts of the country.

Gable roof A double-sloped roof from the ridge to the eaves; the end section appears as an inverted V.

Girder The main structural support beam in a wood-framed floor; a girder supports one end of each joist.

Grade The ground level at the outside walls of a building or elsewhere on a building site.

Joist One of a series of parallel beams used to support floor and ceiling loads, supported in turn by larger beams (girders) or bearing walls.

Lally column A steel tube, filled with concrete, used to support girders and other floor beams.

Lath A building material of wood, metal, gypsum, or insulating board fastened to the frame of a building to act as a plaster base.

Lintel A horizontal structural member that supports the load over an opening such as a door or window.

Nosing The rounded edge of a stair tread that projects over the riser.

Parapet wall The part of a wall that extends above the roofline.

Pigtail (1) A flexible conductor attached to a light fixture that provides a means of connecting the fixture to a circuit. (2) A short length of copper conductor that is attached to the end of an aluminum branch circuit by a special fastener, then fastened to the terminal of a switch or outlet.

Plenum A chamber or large duct above a furnace that serves as a distribution area.

Ply A sheet in a layered construction, such as plywood roofing.

Pointing (repointing) Filling open mortar joints; removing deteriorated mortar between joints of masonry units and replacing of it with new mortar.

Rafter One of a series of inclined structural roof members spanning from an exterior wall to a center ridge beam or ridge board.

Relay An electromechanical switch; a device in which changes in the current flow in one circuit are used to open or close electrical contacts in a second circuit.

Resilient tile A manufactured interior floor covering that is resilient, such as vinyl or vinyl-asbestos tile.

Ridge beam The beam or board placed on edge at the ridge (top) of the roof into which the upper ends of the rafters are fastened.

Ridge vent A low-profile vent system mounted on the roof ridge to help ventilate the area below the roof deck. Works best in combination with soffit vents.

Riser The vertical height of a stair step; also the vertical boards that close the space between the treads of a stairway.

Roof valley The intersection of two roof slopes.

Saddle Also called "cricket." A small saddle-shaped projection on a sloping roof, used to divert water around a chimney.

Sealant Any material or device used to prevent the passage of liquid or gas across a joint or opening.

Sheathing The structural covering, usually wood boards or plywood, over a building's exterior studs or rafters.

Shelter tube Mud-type tube (tunnel) built by termites as a passageway between the ground and the source of food (wood).

Sill plate The lowest member of the house framing resting on top of the foundation wall. Also called mud sill.

Soffit The visible underside of a roof overhang or eaves.

Spalled Chipped and flaked surface of masonry, generally due to frost action.

Stringer (step) One of the enclosed sides of a stair supporting the treads and risers.

Stud One of a series of slender wood or metal vertical structural members placed as supporting elements in walls and partitions.

Subfloor Boards or plywood laid on joists, over which a finished floor is to be laid.

Substrate The supporting material on which the top layer is formed.

Swale A shallow depression in the ground that forms a channel for storm-water drainage.

Therm A quantity of heat equivalent to 100,000 Btu.

Toenail Driving a nail at an angle into the corner of one wood-frame member to penetrate into a second member.

Tread The horizontal board in a stairway on which feet are placed.

Truss A combination of structural members usually arranged in triangular units to form a rigid framework for spanning between load-bearing walls.

Underlayment A material such as plywood or hardboard placed on a subfloor to provide a smooth, even surface for applying the finish.

Weep hole A small opening at the bottom of a retaining wall or the lower section of a masonry veneer facing on a wood-frame exterior wall that permits water to drain.